Praise for
I AM MY FATHER'S DAUGHTER

" ... [Maria Elena Salinas] proves herself to be as irrepressible
on the page as she is on the air."

—*Washington Post Book World*

"Contains irresistible firsthand accounts of interviewing political celebrities, in this case Noriega, Pinochet, Fujimori, and
subcomandante Marcos."

—*Publishers Weekly*

"An inspirational story ..."   —*Library Journal*

I Am
My Father's
Daughter

Living a Life Without Secrets

MARÍA ELENA SALINAS

HARPER

NEW YORK · LONDON · TORONTO · SYDNEY

HARPER

HarperCollins books may be purchased for educational, business, or sales promotional use. For information, please e-mail the Special Markets Department at SPsales@harpercollins.com.

FIRST RAYO PAPERBACK EDITION 2007

Designed by Nicola Ferguson
Photographs courtesy of the author

Library of Congress Cataloging-in-Publication Data is available upon request.
ISBN: 978-0-06-076506-4 (pbk.)
ISBN-10: 0-06-076506-2 (pbk.)

16 17 18 19   RRD   10 9 8 7 6 5 4 3

To the most important people in my life…

Gaby	*Charlie*
Julia	*Isabel*
Bianca	*Tina*
Erica	*Mom*
Cici	*Dad*

Contents

I Am
My Father's
Daughter

ONE

❧

The Box of Secrets

I learned to tell stories from unheard voices, from the nameless Mexicans of my childhood streets. They spoke to me like my mother had spoken, in tender phrases that only hinted of the epic peregrinations they endured. They told me—with some prodding, I confess—about their children, their past loves, their daily hardships, their *machista* husbands, their fairytale dreams. We were kindred souls, linked by more complexities than I could imagine at the time. We had worked side by side when I was fourteen, cutting loose threads from garments in a windowless factory. We met again and again, at random intersections in South Central and East Los Angeles, at the subtitled movies, at Sunday Mass, at the charm school in San Fernando where

I conducted classes, at weekend festivals where mariachis played our favorite songs.

In their stories I found all the key elements of compelling journalism. And when I became a television newswoman, I drew from that well of stories. Of course they seemed different when filtered through the detached lens of the evening news. They often seemed to take on telenovela dimensions. The things that happened to other people, namely my news subjects, always seemed to be more surreal than the things that happened to me. But I, too, had expansive dreams. I wanted to be some kind of self-sufficient businesswoman. But doing what? At first I had wanted to be a fashion designer or a beauty expert. Then, I took marketing courses and developed an interest in the budding Hispanic market, so I dreamed of being an advertising executive. My career path, I decided, would be defined by one requirement: that it never lead me to stagnation or mediocrity. And it never did, although it didn't exactly lead me to the sales office.

THE ROAD detoured at the old, musty newsroom of KMEX, Channel 34, Los Angeles's first and the nation's second Spanish-language TV station. It was an old two-story house converted into a humble, noisy news operation. The air conditioner was always busted, but the Teletype machines worked. In our rip-and-read days, they buzzed constantly, coughing up endless reams of codes, datelines, and bulletins, the hour's urgencies tumbling and curling onto stained linoleum. KMEX was, as then station manager Danny Villanueva called it, "that little Mexican UHF station down the street." But that little station, reflecting the Latino explosion of the 1980s, grew dramatically during my first few years there, and as it did I came to master a new language, one that had been alien to my Spanglish-fluent world. It was the rhythmic and beautifully condensed language of broadcast TV. I was moved by its pace and economy, by the intensity of the breaking news it captured. Here I found a true challenge, a new world to conquer. I wanted to learn everything, so I plunged right in. Soon, I was delivering those familiar Latino stories, the ones I had memorized since childhood, to our viewers, not only during the evening newscast as a reporter and anchor, but also in the daily public affairs show *Los Angeles Ahora* and a weekend entertainment show as host. After a few years and some initial on-air stumbling and bouts of stage fright, I had it down—without a TelePrompTer, for that trusty device came into the picture only much later. I had wanted, more than anything, to make my father proud of me. And he was. In private, he'd counsel me about my new profession. Keep reading, he'd advise, never stop learning.

"Think before you speak," he'd warn. "Read the background of the story that you're covering. You have to be very careful. You have to do this right."

But in public, as my name emerged in the local television world, he'd point to my news reports with pride.

"*Esa es mi hija*," he'd boast.

I was beginning to embrace this new profile of myself as a newswoman, to feel secure in this identity. And then my world was rattled to its core.

My father died. August 6, 1985. That's when I came upon the most daunting story of my life, the one that would challenge my own identity and redefine me.

I GOT the call on a busy news day, in the morning rush for assignments. Days later I stared numbly at my father's coffin from across the room. A kind of force field slammed me against the opposite wall. I couldn't move. Something powerful kept me glued to that wall, away from my father. It felt like gravity. I now understand what the distance between us was about: My father was dead and I didn't know who he was. My father, José Luis Cordero Salinas, was the first family member I had ever lost. He had been in the hospital for six weeks, battling the consequences of a circulatory disease. I had watched him slip in an out of a semicomatose state, gasp for air, and wither away. His death certificate listed about six causes of

4

death. My love for him ran as deeply as the mysteries surrounding his life. Disciplinarian, pacifist, intellectual, undocumented. He had been an enigma, even in our small, tightknit family. He bounced from job to job, enterprise to enterprise. He worked as a realtor, an accountant, a bowling-alley manager, a professor. But it wasn't big bucks or salary bonuses he seemed to be after. Instead, he was driven by a sense of mission and charity. He charged his accounting clients paltry fees. For this, I used to scold him:

"Papi, this is your business. You need to charge them more."

But he'd shake his head.

"No, *m'hija*, they don't make very much money."

Once, he spent all his savings to write and publish a bilingual consumer guide to the real estate market, because he believed the system was unfair to the buyer. I remember another time he took us to a German restaurant. He had never taken us there before, but clearly he was a regular. Everybody there knew him. They spoke to him in German and called him "Professor." He was a reserved man from a well-to-do Mexico City family of opera singers, painters, clerics, and jurists, and yet he had joined the ranks of the working poor in South Central L.A. An enlightened man who had earned several degrees, including a law degree and a master's in philosophy, and commanded at least six languages, he believed in no such education for us, his three U.S.-born daughters. He wanted us to be moral, dedicated wives and mothers, and exclusively so. Besides, he had concluded, the U.S. educational system bred mostly athletes and fortune-seekers. But his devotion to us was such that it wasn't until I reached adulthood that I learned (A) that he was poor, and (B) he had no green card. There were other facts he had concealed from us, the most astonishing of which I would discover several days after the funeral. That's when the Box of Secrets arrived.

I got a call out of the blue from an old family friend. He had something my father had asked him to store at his warehouse. We agreed to meet at the gas station behind KMEX.

"I've had this for a long time," said the friend on that sweltering day as he handed me a square box, about twenty-four inches wide and long. "I have no idea what's inside, but I thought you should have it."

Later that night, at my mother's house, I opened the mystery box. It was filled with books and loose pages, none of them particularly noteworthy. But beneath them I found a tattered, antique leather file. The small, accordion-type file was jammed with personal documents, scraps of our lives: birth and baptismal certificates, report cards, family photographs, official letters, paycheck stubs, rent receipts. Hidden in the compartments were passages of stories he had never told us, letters and documents filled with references to military service, World War II, and alien registration cards. What was this all about? I rifled through the file for more pieces to this puzzle. There were letters, dating from the 1940s, to and from the U.S. War Department, which would later become the U.S. State Department. They alluded to his pacifist convictions, his refusal to go to war, and his subsequent "voluntary deportation." The letters documented a futile, decades-long battle to regain legal entry to the United States. In this campaign, my father cited the "moral education" of his three American daughters as one of the primary reasons why he should be allowed to return to Los Angeles.

One visa request pitched him as an expert of sorts:

> The Mexican National Chamber of Commerce of Garment Manufacturers sees a need for the development of the Mexican garment industry to have a complete technical manual on the subject written in Spanish for distribution among its associates.... [W]e'd deeply appreciate any kind of attention and facilities that you might dispense to Mr. Salinas for this purpose.

But it was in a small church pamphlet that I found the bombshell. It was a commemorative booklet printed for my uncle José Antonio's twenty-fifth anniversary as a priest. In the acknowledgments, my father's younger brother, whom I never met, had written:

> I am grateful to those who were influential in my sacerdotal vocation, including my brother, The Rev. José Luis Cordero Salinas.

I STARED at the name in disbelief. It was my father's name, yes, but "The Reverend"? My father was a priest? I knew he had a brother who was a priest. I knew he had an uncle who was a high-ranking priest. I knew his family had been very religious and conservative in their Roman Catholic faith. But I couldn't reconcile the words before me, that odd salutation before his name—my name. Could it be true? Before I even asked my mother, I felt in my gut that it was. All the subtle clues I had ignored throughout my childhood now rushed my thoughts. He had studied in Rome. He knew Latin. He took long, meditative walks around the park each day. He prized morality above everything else. He had distanced himself, mysteriously so, from his family. I had always wondered why we didn't know most of his siblings, why I hadn't grown up surrounded by cousins, why there were no pictures or stories of his childhood. I always had believed this was because he was a privileged son who married a poor, uneducated woman. I guess I had simply bought into the typical Mexican novela story line, in which the esteemed heir falls for the beautiful, humble señorita and gets disowned by the family. How romantic, I thought, he chose love over social status. But was that really the reason for his estrangement? I knew my mother's life story, its stark and poignant turns. I knew that story very well—or at least I thought I did. But my father was like those unheard, nameless immigrants who came here to reinvent themselves or to blend in with the undocumented populace. Worse. He wouldn't open up, even when prodded.

I closed the file and went to find my mother in the next room. I didn't want to upset her; I knew she was still devastated by my father's death. But I needed to ask her about the church pamphlet I had seen. I didn't want her to slip away to some unreachable distance as my father had. I cut to the chase:

"I found this paper that says my father was a priest."

My mother burst into tears. She was inconsolable.

"I don't know anything."

Gently, I insisted:

"But, Mami, you must have known if he was a priest."

"I don't know…"

"Was it a secret?"

"When I met him, he was an attorney. He was not a priest. That's what he told me."

"What else did he say?"

My mother closed her eyes, as if she were listening to a faint whisper. Then she took a deep breath.

"The only thing your father told me was that he had grown disenchanted with the Church, that he had a big disappointment."

An awkward silence grew between us. Then, she continued:

"But when I met him," she concluded, emphatic, "he was no longer a priest."

My heart ached for her. I felt horrible for reopening a door that had been closed for so many years. But that painful conversation was also a life-changing one. Now we shared a secret, something that for years we would tell no one, not even my sisters. My father's secret became a code for us, a point of entry. From then on, Mami felt she could tell me everything, all those things parents never tell their children. I became her confidante. But I still had many questions. I knew I needed to explore what existed on the other side of that door. This would become my mission, my most difficult assignment, the one that has haunted me for most of my career.

During the next two decades, my job would take me around the globe. From Moscow's Red Square to the jungles of Chiapas, from the streets of Old Havana to the villages of El Salvador, I traveled the world on assignment for the fastest-growing television network in the United States. I interviewed dozens of world leaders, covered wars, natural disasters, diplomatic summits, the death of a princess, the funeral of a pope. I flew by military chopper across flooded villages after Hurricane Mitch ripped up Honduras. I stood vigil by the crumbled home of Salvadoran earthquake survivors as they dug through the rubble for their missing children. I ducked into my car under the aim of a rooftop sniper in a war-battered Iraqi slum. Seven months pregnant, I pounded on the door of a freshly deposed Ecuadoran president, shamelessly begging for an interview. I faced off with Bill Clinton, George W. Bush, and John Kerry. I asked Panama's Manuel Noriega about drug smuggling, Chile's Augusto Pinochet about human-rights violations, and Peru's Alberto Fujimori about corruption.

But facing dictators, wrangling with *comandantes*, and narrating the pope's arrival a dozen times were all walks in the park compared to my personal mission. Pinochet, Fujimori, Noriega—pussycats compared to my father's story. This one's a tiger, not the usual quick-and-dirty parachute job where you jump in, scramble for the story, beam it up in time for the evening news, and move on to the next adventure. This story would never make it on the evening news. I couldn't simply sign off on it.

I would reach the pinnacle of my profession in the macho

world of Spanish-language network news during an era of seismic demographic change. Firsthand, I'd witness the Hispanic population multiply exponentially, from 14 million in the early 1980s to more than 40 million by 2006, and its buying power rocket to over $600 billion. And that immigrant tide would hoist me to an unimaginable place of honor. I became the most recognized Hispanic newswoman in America.

More important, I became a mother of two beautiful girls. If my mission to uncover the truth about my family's past had been intense before my daughters were born, it grew even more so afterward, as a magical bond grew between us. I wanted to keep no secrets from them. I wanted to be close to them, a confidante, as my mother had been to me. Of course, there had been some secrets my mother harbored, but unlike my father she shared them with me before she passed away, in March 1998.

I WAS born in Los Angeles on December 30, 1954. It was my mother's wish that I be born an American citizen, like my older sisters, born four and five years earlier. Many years later I realized she must have wanted to spare me the immigration quandary she had watched my father confront. He had to slip back into the United States, where, legally, he didn't exist. (Of course, he was not the stereotype of the loathed "illegal alien."

He had fair skin, green eyes, and sandy hair.) He rarely left the country, except for an occasional excursion to Tijuana or Ensenada in the years when a California driver's license was good enough for a quick border crossing. She herself had been given a green card after they first moved to Los Angeles as newlyweds in the 1940s. But eventually they moved to Tijuana, where they ran a small factory for some years. What they made—clothing or handbags—I'm not certain. After my birth, we lived in Tijuana for one year. Then we moved to Mexico City, where we lived seven years.

In the Mexican capital, my mother got a job as a seamstress at Esteban Mayo, the famous bridal atelier. I remember going to work with her many times, watching her sew delicate beads onto creamy lace gowns. Well after I was out of diapers, I was still her *consentida*, her thumb-sucking baby, never too far from her apron strings. My mother would give me little swatches of silk to wrap around my thumb at night, in the hope that the soft feel of the fabric would induce sleep and distract me from my addiction. But no such luck. At home, when she measured lengths of fabric for dresses on her enormous sewing table, I'd sit on the edge and watch her. She always made the same request:

"Sing me the song, Malenita."

And I'd break out into our favorite *ranchera*, about a white horse from Guadalajara that brazenly gallops northward.

Este es el corrido del caballo blanco,
Que un día Domingo feliz arrancara,
Iba con la mira de llegar al norte,
Habiendo salido de Guadalajara.

In my memory, that Mexico is a patchwork of colors and aromas. Wafts of hot chocolate on chilly nights. The cheerful *posadas* of Christmastime. My beloved *cazuelitas*, those tiny clay pots I loved to play with. My *estampitas*, the ornate prayer cards I used to buy with my Sunday allowance, the weekly windfall we called *"el Domingo."* I would place the sweet-faced saints before a small altar I scrupulously maintained. Looking back, I realize the saints and the Virgin were well represented and often beseeched in my home. I remember one time my mother made a promise to the Virgin of Guadalupe, and in keeping with her *manda* she had to walk from our house to the Basilica of the Virgin of Guadalupe, miles away. The night before her pilgrimage, I begged to go with her, but she said no. The next day, when I heard her getting ready to leave, I darted out of bed and ran to her:

"Mami, you can't leave me. I have to go with you."

So we walked together for hours until we reached the basilica.

My father spent quite a bit of time away from us, working, they said, in Tijuana. I remember long periods when we didn't see him, when it was just my mom, my older sisters, Isabel and Tina, and me. We were close to two of my mother's sisters and their children, and together we would celebrate holidays and share Sunday meals. But otherwise our family kept to ourselves. My sisters and I went to a private school where we were taught in English. School was different there because the academic year ended in November.

As my father's absences grew longer, our house seemed more and more empty. And, in fact, it was. Something wasn't right. From one day to the next, our furniture disappeared. It was repossessed. We were left with very few possessions. I have a photograph that depicts perfectly this period of our lives: I'm alone in a near-empty room, wearing freshly cut bangs, thanks to my scissor-happy big sister. Around me, there's only one chair, one table, one lamp, and a small porcelain kitten. It was shortly after that picture was taken that we moved to Los Angeles.

My mother had gone to L.A. a few months earlier with my sister Isabel to get settled in. She left Tina and me in the care of my aunt Rosario—Tía Chayo. When it came time for us to leave Mexico, in November of 1963, my aunt dressed me in a blue suit with a white furry collar and put us on a plane bound for Tijuana, where my mother would pick us up. My aunt handed me a small case containing all my documents and told me to guard it with my life. I was afraid to leave my seat the entire flight, worried that something might happen to my papers.

When we arrived at the border, my mother gave me even more instructions:

"Whenever you cross the border, if they ask you anything at all, just say 'American citizen. American citizen.'"

That was one of two phrases my mother learned to say in English. The other one was "apple pie à la mode."

My mother, Luz Tiznado—Lucita to her loved ones—was born to a poor family in El Bajio, a little town of winding, unpaved roads in the northwestern state of Sinaloa. Her father was the night watchman at a sugar mill in a nearby town, El Guayabo. When my mom was eighteen, her mother died, leaving her, the oldest child, to care for her seven siblings and her father. She met my father a decade later when she moved to Mexico City and took a job as a dentist's assistant. My father was a friend of the dentist, she told me. She was a beautiful woman in her late twenties when she met him. But it wasn't until late in her life that I would learn the painful secret that had prompted her move to the capital.

TWO

❧

Miss Mexico of L.A.

Alfonso Tirado was a charming man whose sterling name was known far and wide across the state of Sinaloa. His father owned the sugar mill in El Guayabo where my maternal grandfather worked, and, like many young girls in the region who swooned over Alfonso's good looks and strapping frame, my mother knew of his reputation as a lady's man. Then again, how could he not be? He was a privileged son with dreams of provincial leadership. Plus, he rode a beautiful horse.

Many years later, my mother would recount, in moments of reverie, how in those days such men would gallop through the villages at night, scooping up the loveliest young ladies and spiriting them away. The villagers were left to spin the story of the vanished maidens.

"*Se la llevaron al río*," they'd conclude, as if the disappearance was the doing of an entire army. They took her to the river.

By modern-day standards, of course, such an act would be tantamount to aggravated kidnaping and felony Godknows-what-else. But as my mother would tell the story of her own vanishing from El Guayabo, upon the horse of Alfonso Tirado, she described no such crime. No, it was a love story she told, one so passionate and pure that even six decades later it would bring a glow to her face. It was the kind of love story that inspired all the great *boleros*, even the ones that make you cry. As she recounted, they were deeply in love, although the odds against their relationship seemed insurmountable. He was rich; she was poor. That fact alone guaranteed she would never be his wife, for marriages across class lines were unheard of. And, by her own admission, she wasn't even his only girlfriend. But, she'd add with a gleam, she was his favorite. No doubt she was the prettiest girl in town, a green-eyed beauty with auburn hair and Indian cheekbones. The sepia-tinged photos of her youth attest to that.

Despite her humble roots, my mother always carried herself with elegance. Even though she couldn't afford to shop at fancy stores, she was a stylish woman with an eye for detail, always matching her shoes with her purse. She could make the blue-light special at JC Penney look like high-end couture. In my eyes, she was this tall, willowy, youthful woman, even though she gave birth to me when she was forty-two. When I grew older, I realized she wasn't as tall as I had imagined, for I towered over her. Still, I took it as confirmation of her impeccable posture and poise.

Certainly, it was this radiance that caught the eye of Alfonso Tirado. He furnished a tiny apartment for her in the state capital, Mazatlan, where he would become the mayor. My mother lived there with her younger sister, Conchita, a sprightly, freckle-faced redhead who had taken the name of her favorite singer because she loathed her given name, Petra. Like my mother, Conchita/Petra didn't care for their father's last name, Tiznado. It didn't have the poetic ring of their mother's family name, Lizárraga, a respected name in the state of Sinaloa, where one of the most famous musical groups, La Banda del Recodo, was founded by Lizárragas. But Tiznado seemed to derive from *"tizne,"* which means soot. So Petra Tizne became Concepción "Conchita" Tirado. For a while, my mother used that last name as well, for it was also the name of her boyfriend, Alfonso.

My mother's love would visit every day. This is what I knew for many years, as Mami would recount the story, a beautiful love story, in some detail throughout my life. But what I didn't know was the story's tragic ending. And there was one startling fact she kept secret until the week of my father's funeral. It came to light, thanks to her sister, who, upon seeing my mother so devastated by my father's death, confided to me:

"This is the second time I've had to console your mother for losing a loved one."

"When was the first time?" I asked.

Her reply left me speechless. I waited for the funeral to pass and then invited my mother to Santa Barbara for a few days. We drove along the jagged coast that linked her present and her past, against the same Pacific breezes that caressed her beloved Mazatlan. At last away from the rest of the family, we were once again alone, confidantes. I told her I knew her long-kept secret:

She'd had a baby with Alfonso Tirado.

My mother nodded sadly.

"Yes, I had a little daughter who passed away."

Her name was María de Los Angeles. María of the angels. She was about three years old when she died. Knowing of her existence cleared up many things for me. The revelation that there was a baby before my sisters and me explained the strange wording I once detected on my birth certificate. On the line that asked how many children my mother had given birth to previously, she had answered "three." Three? When I asked her why she had not written "two," considering she had only my two sisters and me, she sent me off:

"Go ask your father."

And, of course, when I asked my father, he only echoed her reply:

"Go ask your mother."

So, not only was there a fourth sister, but one who had died long before we were born. Still, my mother and I decided we would keep this story between us and years passed before we told my sisters.

As she told the story, Alfonso Tirado had married someone else, a high-society girl with whom he had a child. But my mother, his Lucita, was the love of his life. She said he proved to be a doting father who spent a lot of time with her and the baby. But one day, while he was away, the child fell ill. She lost strength very quickly. My mother was desperate. She tried all kinds of home remedies. The hospital was far away and she had no transportation and no means of communication. She had no way of getting an urgent message to her love.

By the time he got there, the baby had died. He was devastated, so much so that he grew even closer to my mother. They remained together, even as he pursued his political dreams. As she told the story, Tirado decided to run for governor of Sinaloa. But those dreams would be short-lived. On the campaign trail, his rival ambushed and killed him.

My mother, still grieving for their baby, could no longer stand the pain of living in Mazatlan. So she left for Mexico

City, hoping to start a new life. It was there that she met my father. When she was about thirty, they married.

But she never forgot her first love. On the fiftieth anniversary of his assassination, old friends in Mazatlan sent her newspaper clipping. The local press had published a commemorative piece. She rushed to show me the article.

"Look, he was my boyfriend when I was young."

After they were married, my parents spent six childless years, moving between Los Angeles and Tijuana. For years, the details of that era were sketchy at best. Their love story lacked the tidy data of most other marriages. There were no details of the nuptials, no photos, only a simple wedding band. It would take me many years to find out why that was. I had always believed that whatever mystery shrouded their relationship, it had to do with the fact that they came from vastly different backgrounds. My father came from a very refined family. In Mexico, they use the word *abolengo* to describe this class, which, though not necessarily wealthy, is certainly distinguished. And, in the case of my father's family, seemingly intolerant. How else could one explain his distance from them?

I never fully understood why my father's name changed when he crossed the border into the United States. In Mexico, his last name was Cordero Salinas, the dominant being Cordero, the paternal surname. In this country, it became

Salinas, the Cordero reduced to a "C," a seldom-used middle initial.

Perhaps this was the doing of a clueless immigration official unfamiliar with the Latin American tradition of extended last names. Or perhaps, as my father once told me, he simply identified more closely with his mother's family. In truth, I had no idea.

Whatever the case, we were all Salinases when we returned to South Central Los Angeles from Mexico City in 1963. It was fall, what would have been the end of my third grade in Mexico. The overlap caused me to repeat the grade. Another reason they didn't bump me up to fourth, despite my good grades, was because the third-grade teacher spoke a little Spanish and I couldn't communicate very well in English, even though I had taken English classes in my Mexico City school. There were only a few Hispanics at the time in the predominantly African-American school. It took me a while to talk to anyone other than my teacher. My sisters went to a different school, where they were in English-as-a-second-language classes. But at my school, I had to learn English straight up because no one there, except for my hardly fluent teacher, spoke my language. So I learned English in six months and eventually lost my accent. (Of course, it would creep back decades later when I went to work for Spanish-language TV.)

I used to walk home by myself. In fact, I remember spending a lot of time by myself. I didn't participate in after-school sports or activities, not in grade school. In fact, I was a latch-key kid at age nine. I used to come home to an empty apartment and start dinner, straighten up, iron clothes, and do my homework until my mother and sisters arrived, hours later. Then again, I never felt afraid or alone. Maybe that's because the presence of my parents loomed from the moment I stepped into the tiny kitchen. There, posted on the wall, were my father's typewritten rules. In military fashion, he spelled out our daily schedule, from the hour we rose at 6:45 A.M. to the hour we went to bed at 10:00 P.M.:

MEMORANDA

+ Return things to their proper place after usage. For example, place towels on their racks, clean clothes in the closet, soiled clothes in the hampers.
+ Maintain a clean house, especially the bathroom, living room, dining room, and kitchen.
+ Wash and hang clothes twice a week.
+ Take out milk bottles on Mondays and Fridays.

CONDUCT

+ Entrust oneself to God upon rising and retiring. Attend mass on Sunday. Say confession and take Holy Communion with frequency.

- Say please and thank you.
- Say hello and goodbye.
- Eat and dress with decency and urbanity.
- Do not walk barefoot or chew gum.
- Obey promptly and with a good attitude.
- Do not utter vulgarities or insults.
- Avoid idle, useless, and dangerous acquaintances.
- Show eagerness to learn and avoid idleness.
- Always ask permission to go out. Advise of your whereabouts to avoid parental worry.
- Avoid bickering and screaming. Speak in normal tones.
- Do not use the telephone for more than 10 minutes at a time.
- The sisters should accompany one another. In the absence of the parents, the younger should obey the older.
- Have trust in your parents and seek their counsel.
- Money, outings, and diversions are rewarded based upon merit and good conduct.

My father's strict codes should have been a clue to his pious past, but I never suspected they were more than the manifesto of a disciplinarian parent. To say my father was strict would be an understatement. Sleep-overs? Forget it. Even into adulthood, I would not sleep over anyone's house, for fear of my father's disapproval. But he had no reservations about my walking to and from school, or all over our South Central neighborhood to sell chocolate. Now, as a mother who wouldn't dream of letting my daughters walk one mani-cured block in our tony neighborhood by themselves, I find this to be a startling fact. We lived within the tense circumfer-ence of Watts, smack in the middle of the 1960s, an era when deeply rooted frustrations over inadequate housing, soaring joblessness, crumbling schools, and rampant discrimination threatened to boil over—and did.

To my father, South Central Los Angeles was like any-place else, a place to walk and meditate amid the sounds and scents of the city. He carried his Bible on long, serene walks past all the neighboring landmarks, the sports arena across the street, the park, the Coliseum, the museums, the sprawl-ing campus of the University of Southern California.

As for me, I can recall just one scary instance during one walk to school, shortly after we arrived from Mexico. I noticed a strange man in a slow-moving car, staring at me as he fol-lowed my steps. Panicked, I scurried to a nearby house and knocked on the door. I didn't know how to explain what the man was doing, so I just blurted:

"He see me! He see me!"

But the people at that house ignored me. Luckily, I spotted a milk truck making its rounds and ran up to the driver.

"He see me!" I cried.

The milkman signaled for me to hop in. He drove me to school in his little truck and escorted me to my classroom. As I sobbed, he told my teacher what had happened. Fortunately, it never happened again.

There were other high-anxiety moments a couple of years later, but they were caused by a general sense of hopelessness in the surrounding areas. The Watts riots, which gripped the city for several days in August 1965, exploded across my neighborhood streets. I remember taking the bus home to our apartment on the corner of Figueroa and Santa Barbara streets and then having to run the remaining half-block to the front door because we knew armed rioters were hiding in the park across the street, waiting for the first shadows of dusk. In the apartment, my mother, fearing stray bullets, made us sleep on the floor and duck if we walked near the window. We could hear the gunfire and militants outside on Santa Barbara, a street that was later renamed Martin Luther King.

There was no shortage of excitement inside the apartment, either. We lived in a fourplex, in a ground-level unit crawling with the most obstinate cockroaches in L.A. No amount of roach spray deterred them from their mission to take over the house, which we took to calling La Casa de las

Cucarachas. God forbid you got up to fetch a glass of water in the middle of the night... *Whooooosh!* They'd scamper across the floor ahead of you. They were everywhere, in the kitchen cabinets, the closets, the bathroom, all of them foot soldiers from a fierce encampment in the attic. It didn't help that I slept in a closet. In the cramped two-bedroom apartment, my parents had set up a cot for me in our "walk-in" closet, a tiny space with an angled roof tucked beneath the stairs that led to the apartment unit on the second floor. Everything resounded in that closet. My pillow rested against a wall that separated my space from the laundry room. At night, when my mother washed clothes, ghostly *woooo* noises reverberated through the wall. Thunder, engine roars, and even fireworks on the Fourth of July echoed like bombs in there. I think that's why for years I was deathly afraid of thunder and lightning.

I'm not sure how sleeping in that closet affected me, but I can say that I never once took a single square foot of space for granted after that. Many years later, when I bought my first house, my aunt reminded me of my old quarters.

"Do you remember when you used to sleep in the closet? Now you have three bedrooms and you can sleep in any one you want."

And, trust me, I did. I even slept on my living-room floor, sprawled beside my fireplace. One more thing I learned from those years in La Casa de las Cucarachas: I knew I always wanted to have space for family. So each time I bought a home, I made sure there was an extra room for my mother. And for the cucarachas, proper Roach Motels.

AT SCHOOL, although I got only mediocre grades, my teachers were convinced that those grades did not reflect my potential. They believed I could do better and sometimes suggested that I was too "sociable." I still find that hard to believe, considering how painfully shy I felt throughout my childhood. But I do concede that I wanted to appear more grownup than I was, especially in the eighth grade, in that giddy stretch before high school. I guess this was because I had two older sisters who wore exotic, grown-up garments, like nylons. My legs were too skinny for nylons, but that didn't stop me from pulling them on every morning beneath my school clothes. It was my daily exercise in big-girl optimism, one that withered as the afternoon approached and the stockings sagged hopelessly around my knees.

Ill-fitting hose aside, I was in love with fashion. I wanted to wear the latest designs, create new looks, master all the tricks. In high school I joined the sewing club. Soon enough, I was putting together fashion shows to display my creations—and my budding emcee talents. I narrated as my models—namely, my classmates and my three-year-old niece, Cici—showed off my outfits. As I watched them sashay along the stage, I dreamed of breaking into the fashion industry, a glamorous world, I imagined. As it turned out, my first job in fashion—my first job, period—was as glamorous as my old nylons. At fourteen, I took a summer job at the clothing factory where my mother worked as a seamstress. My assignment was to cut dangling threads from the finished garments. They gave me a pair of tiny scissors to snip the threads. It was the final step before the pieces went into plastic garment bags for shipping. It was easy enough, but I was terrified I'd miss and ruin the clothing. I was so nervous on my first day, I ran to the bathroom and vomited. Once school started again, I had to quit the factory. I took an after-school job serving food at Clifton's Cafeteria on Broadway.

Later, I went to work at a movie theater that was part of the Metropolitan Theaters chain of cinemas that showed Spanish-language and Spanish-subtitled films. I started by selling popcorn and scrubbing floors and eventually worked my way up to ticket sales. I was thrilled at the promotion—I was inching upward, improving, something my mother taught me to strive for. I would watch her bring piecework home from the factory and plow through it while we watched TV. She was so fast that she reasoned it was better for her to be paid by the piece than by the hour. So she always found a way

of moving forward, inching upward. Her work ethic was re-
markable, as were her money-management skills. What still
amazes me today is how, no matter how cramped we were or
how humbly we lived, I never once felt poor. I don't remember
feeling we lacked anything, particularly not my mother's presence, in spite of her long hours of work. Nothing was missing.
Everything was there, thanks to her. She was my idol. I wanted
to be just like her, loving, tolerant, graceful, hardworking, fru-
gal. I scrambled to balance schoolwork and my after-school
job. At Christmastime, I'd take on a second job at a department store to boost my earnings, which I would divide into
three portions: one-third went to my mother for rent, another
went to Our Lady of Loretto Catholic school for tuition and
my expenses, and the remaining third went into my savings
account. I had little time for anything but school and work. If
I was going to be a fashion designer, I thought, I didn't have
time to dally. That was one thing my father's rules drilled into
my head: NO IDLENESS ALLOWED. Thinking back,
those rules were a great thing. They certainly kept me out of
trouble, unlike my best friend, Rita, who got pregnant at age
fourteen. We were snotty-nosed kids, and there we were at
her wedding. I was her maid of honor and later her daughter's
godmother—at fourteen! But it wasn't such a rarity, really.
Several of my girlfriends got pregnant around the same time.
I think my father must have thanked his lucky stars that I was

a prude, although his two older daughters also got pregnant as teenagers. As it turned out, I didn't leave my parents' house until I got married. That doesn't mean, of course, that I didn't have boyfriends. When I was sixteen, I actually got engaged, to a guy named Ray. Well, "engaged" is only the technical word for what happened between us, a couple of sweaty-palmed kids who never went beyond a timid kiss. In Spanish you'd call it *"una relación de manitas sudadas."* Let's just say my engagement ring came attached to a huge teddy bear. And I liked the bear better than the ring. That summer, Ray and I went to Acapulco with my mom and something amazing happened: the world got bigger. There were beautiful beaches and swaying palm trees, and, more importantly, lots of other guys. And I was engaged? Not for long. At the beach one day, I broke the news:

"I don't think I want to get married anymore, okay?"

Poor Ray didn't take it well. He got up and hurled the engagement ring into the ocean. So I always believed there was a fish out there, off the coast of Mexico, with a tiny diamond in his belly.

I GRADUATED from high school in 1973, still with big dreams of becoming a fashion designer. But here was my problem: I had asked several school advisers where I could go to study fashion and they all told me the same thing—trade school. Not what I wanted to hear. I wanted to hear "USC" or "UCLA." The thought of trade school was depressing, although the world of fashion and beauty still held a magical allure. I resolved to break into that world.

I didn't realize that my first opportunity would come in the form of a beauty pageant. Beauty pageant? Me? I wanted no part of it. But at the urging of my sisters and family friends, I entered the Miss Mexico of Los Angeles beauty pageant. The name on my sash didn't have a Mexican ring to it. "Miss Miller," it said. My sponsor, a dear family friend and promoter named Tony de Marco, handled public relations for Miller Brewing Company. He was a pretty well-known guy in the community, as he also was baseball star Fernando Valenzuela's manager. I won second runner-up in that pageant. More important, I got to show off my designing skills. For the talent contest, I put together a mini–fashion show, showing off original outfits that symbolized the four seasons. After that, I entered the "Reina de la Hispanidad" pageant twice, and both times I won first runner-up. That was the end of the beauty-pageant circuit for me. That era won me no crowns, but it did give me something more important: a jolt of self-confidence. On the stage, little by little, I lost the shyness that had gripped me in school.

Losing my shyness was a good thing, but, to my parents'

dismay, it also brought out the rebel in me. I grew restless, chafing at my father's unbending rules. At nineteen, I wanted to be on my own. On a whim, I took off for Mexico. I went to live with my cousins, Lila and Hilda Deneken, and my aunt, Conchita. I spent six months adrift in the capital. I couldn't get a job because my Spanish was rusty by Mexican standards and I didn't have work papers. So I did nothing. Six months later, my sister Tina's husband, Jorge Rossi, came to the city on a business trip and delivered this blow:

"Your parents are suffering very much. Your mother says she's going to die if you don't come home."

I didn't want to go, but his words tugged at me. I went home.

I went to work for a garment manufacturer and enrolled in marketing and merchandising classes at East Los Angeles College. I also returned to what I knew best at the time: the beauty-contest circuit, not as a contestant but as an organizer. I teamed up with Tony de Marco and began emceeing pageants and special community events, like the Cinco de Mayo parade in Disneyland. Through these events, I got a glimpse into Los Angeles's vibrant community of grassroots, nonprofit organizations. Years later, those groups would play an important role in my budding career as a TV reporter and anchor, opening doors into the heart of Mexican-American L.A. But being a broadcast journalist was the farthest thing from my mind at the time. I didn't want to be Barbara Walters. I wanted to be Margarita O'Farrill.

MARGARITA O'FARRILL, inventive business-woman and charm-school doyenne, was a community icon and the mother of my good friend, Maggie. She ran a successful academy that taught classes in modeling, fashion, hygiene, culture, and etiquette. She was beautiful and elegant, a walking affirmation, always ready with inspirational nuggets I'd savor for years.

"No hay mujeres feas," she'd say, "sólo mujeres que no se arreglan."

Which means: There are no ugly women, only women who don't fix themselves up.

This was her philosophy. If you look good, you'll feel good about yourself. And when you feel good about yourself you can do anything. Sometimes we don't dare seek what we really want because we think we'll never get it. This is what she used to preach: Eliminate the obstacles in your life that make you fail, be it negativity, a particular person, whatever it may be. She believed beauty came from the inside, from a well of self-confidence every woman could tap, no matter her social circumstances. And she wanted me to work for her, giving makeup lessons at one of her charm schools. At age twenty-two, I felt honored. I fancied myself a makeup expert and often made guinea pigs of my sisters, their friends, and even the baby-sitters.

My classes were a hit. Margarita was so pleased she offered me a huge promotion: she asked me run one of her schools and launch a program for adolescent girls. I jumped at the opportunity. We opened up a school in the San Fernando Valley and started giving youth classes on everything from proper manners to cleanliness. I would visit area parochial schools and negotiate with the priests:

"I'll put together a fashion show for your next school dance if you allow me to go into your classrooms and distribute charm-school flyers to the girls."

Our student population quickly boomed. Not yet twenty-three, I felt like a successful businesswoman.

But perhaps my greatest sense of accomplishment came when I began teaching classes to a group of immigrant women, mostly poor and illiterate, who had discovered our school. They would come during the day while their husbands were at work. Charm school was the secret they kept from their *machista*, controlling men. Their husbands, they said, would never allow them to learn something new, because they didn't want their women to be smarter than they were. It was clear the men stepped all over them, treated them like worthless objects, in some cases beat them. But the women's greatest obstacle was not their husbands' disapproval, it was their own fear of knowing too much.

"What good does is it do me to learn all of this when my husband doesn't know what to do in a social setting?" they'd tell me at first. "It'll only make him feel uncomfortable."

But as time passed that attitude faded. I watched the women become more assertive, more outspoken and sociable. By the time graduation rolled around, they took the stage like supermodels, beaming before the sparse gathering. Their unwitting husbands were at work.

Those classes also proved to be a confidence-booster for me. Their endless questions forced me to work on my extemporaneous skills, to articulate my thoughts succinctly before an inquisitive audience. I learned to think on my feet, quick and clean. So, in a sense, Margarita was responsible for preparing me for a career as an anchor. She was also responsible for another, less successful, chapter in my life. She introduced me to the man who would become my first husband, a radio disc jockey named Eduardo Distell.

I was drawn to his charm and great sense of humor. And, very quickly, we forged quite a team on the special-events circuit, where Eduardo often played the honey-voiced announcer. Through him, I learned all about radio—the mechanics, the culture, the behind-the-scenes drama. I visited him so often at the station, KLVE (K-LOVE, Radio

Amor), that it felt like home. It wasn't long before I, too, started working in radio, at Radio Express, as a deejay, spinning romantic Mexican *boleros* on my own show and reading wire copy on newsbreaks. I enjoyed the business of radio, but the deejay gig didn't hold my interest for very long. I was more curious about the larger picture, particularly about what was going on in the station's sales and marketing department. I felt drawn to the growing world of Hispanic advertising and its early possibilities. That's where I belonged, I thought.

Unfortunately my marriage to Eduardo was a dead end. It lasted only a year and a half. And suddenly the radio station seemed smaller than ever, too small for the two of us. I wanted to get as far away from it as possible. But I was torn—I wanted to remain close to the business. Radio had opened up a new, mind-boggling world for me, a world revolving around an exciting, ever-growing market, a population that spoke my language and understood the nuances of my culture. This was my future, I could feel it. I wasn't ready to leave it behind, no matter how badly I needed distance from Eduardo.

So when I heard that the leading Spanish-language television station, KMEX, had an opening for a host on its afternoon public-affairs show, *Los Angeles Ahora*, I saw my chance to branch out. I thought it would be an ideal part-time job to support me while I explored my main area of interest—advertising. A public-affairs show could provide a showcase for those grassroots community groups I was becoming so familiar with. And while it was a welcome challenge, it wasn't such a time-consuming job that it would force me to quit my radio work. Or so I assumed. In any case, when the station held a casting for fluently bilingual hosts, I was eager to audition.

Surprisingly, I got a callback. I was hired. That was the great news. But moments later, I learned the bad news: not only would I have to leave my radio work, I would also have to assume a slew of responsibilities I'd never bargained for. Yes, I would be the new host of the public-affairs show. But here was the shocker: I would also be a reporter and anchor for KMEX's evening newscast.

Los Angeles Ahora, indeed.

THREE

❧

Channel 30-WHAT?

The first day of my new TV career began with a barrage of reality checks. This may sound like a TV cliché, but my hair was "all wrong." It poufed out in carefree waves. It was good radio hair, just like my work outfits were good radio outfits, comfortable and carefree. But my fashion sense, keen as I thought it was, was tailored to another world. I was a voice, not a face and figure. But that was the problem: the image in the mirror all but screamed "FM." I didn't look anything like TV people are supposed to look. Then again, what did I know about TV? I had done a couple of local commercials, but that was about it. So I dressed again and primped and blow-dried, all the while trying to fathom the three-jobs-in-one situation that awaited me at KMEX, Channel 34. The public-affairs show alone was a daunting proposition, without the added responsibilities of reporting and anchoring for the prime-time newscast.

María Elena Salinas reportando desde los estudios de... KMEX, Los Angeles.

The jazzy sign-off rang in my head as I smoothed out my curls. I liked the ring of it.

... reportando desde Los Feliz, desde el Este de Los Angeles, desde San Fernando, María Elena Salinas, Noticiero 34...

I rolled my shoulders back and cleared my throat, then tried the sign-off on for size. And then I gasped. Nothing came out. No sonorous TV voice. No sound at all. I had no voice. Laryngitis. My frazzled nerves had silenced the new voice of KMEX.

It was April 9, 1981, in the City of Angels, a crisp, clear day. Outside my window, an immigrant town, a largely over-looked and undocumented population, churned to life, fueling a thriving underground economy. Their "foreign" language seeped into every crevice of Los Angeles, drifting across the hills and over the valley, past meticulous lawns, perfectly tended flower beds, and kitchens fragrant with the aromas of Oaxaca, Sonora, and Guanajuato, even San Salvador and Chichicastenango. The merchants of East L.A.'s most vener-able thoroughfares, Olvera and Alameda streets, Brooklyn Avenue, Whittier Boulevard, snapped open their shutters to their earliest customers. And their bustle would echo only within the confines of the community and on the airwaves of Channel 34. Oh, yes, there was the occasional sound bite on KCBS, KABC, or KCLA, gathered by whatever unlucky soul was stuck with the dread "taco beat" that day. But the people of this community, mostly ignored by traditional news outlets, were the domain and the pride of KMEX, the tiny UHF station they called "*nuestro Canal 34.*" It was the largest in a cluster of stations owned and operated by the Spanish International Network, SIN, which years later would become Univision. "Network" is an ambitious term for that initial string of humble affiliates in L.A., San Antonio, Miami, New York, outlets for Mexican soaps and an imported international newscast, *24 Horas.* The network was born in the early 1960s, with four stations connected by satellite. By 1980, it had picked up seventy-six affiliates. Five years later, SIN had 364 affiliates, a number we considered mind-boggling at the time. (Who could have predicted that the number of Univision affiliates would mushroom to more than 1,500 today?) Noting the staggering numbers in 1985, the *Christian Science Monitor* reported a projected $55 million for SIN's advertising revenues that year, a leap from $15 million in 1980.

When I started at KMEX in 1981, our operation was so low-budget, we barely registered on the mainstream media radar. For years, baffled newsmakers and sources greeted us as

"Channel Thirty-*WHAT*?" We'd explain that we broadcast in Spanish and that we wanted to interview a spokesperson who could communicate in the language of our viewers. Amazingly, they'd cart out the janitor or some other low-level employee. Never mind that at the time Los Angeles had more Spanish-speaking households than any other media market in the nation. And they did not tune in for some stilted version of the nightly news. The people we featured on our screen were not limited to Latinos in handcuffs, or the subjects of police reports. Our viewers tuned in because they saw themselves reflected in the faces on their TVs, and they understood the language. I knew this going into my new job, and this was what enticed me. I wanted to report for them, and from their midst.

But that first day when I showed up in the newsroom, I felt pretty useless as a reporter, much less a talk-show host and anchor. I was eager to meet the other reporters; then I found out there was only one reporter on staff, and I would be the "other reporter." Still, I watched this reporter, Mario Lechuga, an older man with a Pancho Villa mustache, assemble a story, patching together narration and sound bites. I made use of my speechless first days to learn my way around. My voice didn't come back for two weeks. When it did, I ventured out with a couple of weather reports. *I can do this*, I thought. But soon enough, the new job flurries became a blizzard.

Slipping into the host chair on *Los Angeles Ahora*, the station's community affairs program, which aired live Monday through Friday at three o'clock, proved to be my first real challenge. Public affairs was Channel 34's forte and what set it apart from the "mainstream" stations, as it reached into the households of L.A.'s Latinos. Just a few years later, tapping into the heart of this fiercely loyal viewership would bring KMEX ratings that not only rivaled its English-language counterparts but often surpassed them. Here was a trust built one public-service hour at a time. And nurturing the relationship between the station and the community would be one of my most cherished responsibilities, one that I would carry out for six years.

The news side was another story. I had to report at least two stories a day in addition to preparing my script for the newscast. The camera on the news set petrified me. It stared at me like an X-ray machine. I was too embarrassed to look up from my script. Yes, we had a TelePrompTer, but nobody knew how to work it. So I just read from my script, once in a while peering up for emphasis.

My co-anchors were Eduardo Quesada, a veteran newsman, and Paco Calderon, a police officer turned broadcaster. Paco was also the news director, a no-nonsense type with little patience for stories that ran too long. He had a brutal way of editing a rookie's reports. God forbid I'd give him thirty seconds more than what he had budgeted—he'd simply lop them off from the end. As a result, many of my early stories had a curious way of screeching to a halt, mid-climax.

"But I have so much to say," I'd vent.

I waited in vain for useful tips or words of journalistic wisdom. Instead, Paco simply told me to just say it all quicker.

"*¡Si te pasas de dos minutos, te corto!*" was his usual warning.

How could I keep every story to two minutes? Needless to say, I was pretty frustrated with the entire scenario, but then again, how could I expect pearls of wisdom from a guy who'd put a novice like me on the anchor desk? God bless Paco Calderon and his SWAT-like approach to the evening news. By throwing me into the madness, he showed me the most thrilling part of the news business, the daily rush against the clock to capture, decipher, and deliver the day's happenings to one's audience. Breaking news, I learned, was an unforgiving taskmaster, for it never waits for you to catch your breath. What a ride, I thought after my first couple of weeks on the job. I was hooked. I wanted to learn everything I could about journalism, so I enrolled in a broadcast journalism program at UCLA Extension. There was one thing I learned growing up: no matter what you do in life, try to be the best you can be. If you're going to be a seamstress, be the best seamstress. If you're going to be a janitor, be the best janitor. So I told myself that if I wanted to be a broadcast journalist, I was going to do it right. I became obsessed with news. I watched every newscast I could, devoured newspapers and newsmagazines. Newly divorced, I lived alone in a one-room apartment. I cleared my life of distractions. I lived, breathed, and ate news. I joined the California Chicano News Media Association and became active in its Spanish-language media committee. I worked alongside José Lozano, whose family owned L.A.'s *La Opinión* newspaper, to organize community forums. We would teach nonprofits how best to deal with the media and get their message out.

This work dovetailed nicely with my public-affairs show, which promoted health fairs, job opportunities, fundraisers, and cultural events. The only topic that made me a little antsy was the periodic antismoking messages brought to us by reps from the American Heart Association, who insisted on showing graphic pictures of blackened lungs and delivering dire warnings about cancer. They were nice enough people, but their campaign had the opposite effect on me. During the breaks, I'd excuse myself and go outside and smoke—and smoke and smoke. It took me years to heed their warning and kick the habit.

Press conferences made me nearly as nervous. For years, I was too shy to ask a question. Instead, I'd take notes on the other reporters there. I would sit in the back of the L.A. Press Club and observe this remarkably tall, blond reporter from KCBS as she worked the room and asked great questions. Her name was Paula Zahn. Each night, I would tune in to her newscast to see how she had put together her story. (The KCBS anchor, by the way, was Connie Chung.) I'd compare her stories to mine, and her questions to those streaming through my head. After a while, I realized my questions weren't as dumb as I had thought they were. This simple realization did wonders for my confidence.

A few months after I started working at KMEX, we got a new boss. Pete Moraga was a veteran newsman and the best thing that happened to me as a novice journalist. He was smart, experienced, and a keen judge of talent. His first order of business involved me.

"The first thing I'm going to do is fire that anchor," he told himself when he caught sight of me reading into my script.

I adored him. *Que en paz descanse.*

No, Pete didn't fire me, but he did yank me off the anchor desk. He realized that I had been tossed into that job without experience. He told me to focus on being a good reporter. And, with Pete as my mentor, I did. He had worked at the Voice of America, as a U.S. press attaché in Mexico and Peru, and as a radio commentator in Los Angeles. And now he would be my first true journalism professor.

I'd bombard him with questions about story structure and news in general.

"I'm not sure I understand what the story is," I'd say time and time again.

"Okay, okay, settle down," he'd say. "Now tell me what they said at the press conference." And I would recount the gist of the story, to which he'd inevitably retort:

"Well, say it just like that."

"Like what?"

"What you just told me—that's your story."

Pete was a great news director, but he was tough as nails. When he sensed his staff wasn't devoting the proper energy to writing our own stories in Spanish, he canceled the Notimex Mexican wire services, the only Spanish-language wire we had.

"I don't want you to rip and read. I want you to report the news, interpret the news, tell the story."

About a year after he got to KMEX, Pete put me back on the anchor desk. I was ready to look up from my script.

I must have done something right, because a couple of years later, I was approached by the mighty KCBS. The station was looking for a Hispanic reporter to cover Hispanic issues. Many Latinos trying to make it in the mainstream might have been insulted by the so-called taco beat. But I felt honored at the thought that one of the big network affiliates in Los Angeles was finally going to cover the city's fastest-growing minority in a meaningful way. So I taped an audition for them. But in the end I didn't get the job. The news director liked me, but when it came time for the general manager to weigh in with the decision, he passed. His conclusion, according to my attorney, who was negotiating the deal: I didn't look ethnic enough, and I sounded *too* ethnic.

"Her accent would insult our general audience," I was told he said.

The insult was all mine. I took pride in being fluently bilingual. As for my "nonethnic" looks, I can only say I look like my Mexican parents. The station hired a young woman with long black hair and brown eyes. She spoke English without an accent—whether or not she spoke Spanish didn't matter. The curious thing about that episode is this: one of the anchors at KCBS at the time spoke with a hearty British accent. I never understood why one accent could be considered to be sophisticated and charming, while another was taken as offensive. So much for my short-lived crossover dreams.

In the long run, though, I lucked out. I went on to travel the world and cover historic events, superpower summits in Moscow and Washington, elections in Latin America, natural disasters, wars, papal visits. More important, I got a first-hand look at one of the most dynamic stories of my generation, the sweeping growth of a Latino population that would change the face and, yes, the accent of America. The immigrant tide from Latin America would prove to be an unstoppable force. Not only would it transform American cities, it would transform the marketplace, and with it the communications industry.

When I began my career in television, the U.S. Hispanic population numbered about 14 million. Today there are more than 40 million, with a combined buying power of over $600 billion. At times, watching this explosion from the inside has been nothing short of mind-numbing. The numbers carried with them echoes of lands I came to know well in my travels throughout Latin America, as well as phrases and dialects that would open up the American vernacular.

But here was the frustrating part: their soaring numbers did not translate to proper representation for Hispanics in the larger society. A hostile and powerful minority of anti-immigrant activists saw to that, relentlessly lashing out at the Hispanic surge. For their part, too many Latinos seemed clueless as to their potential for creating a political force. Still, the numbers were there and they were tuning in to Spanish-language television.

I can't tell you how many times I heard the would-be prophets of the broadcast industry declare that this was just a fleeting trend and that there was no future in Spanish TV. "Hispanics will assimilate eventually and Spanish TV will just die," they'd say, adding: "You really should make the transition to English-language news."

Well, they were right and they were wrong. Hispanics did assimilate, but for them assimilation didn't mean leaving behind their language and cultural nuance. On the contrary. Even their American-born children catch that "roots" fever sooner or later and make the plunge into their parents' heritage, even if only for a hot song, an enchilada binge, or a great guayabera shirt. It's inevitable.

I dismissed the naysayers and stayed. So as my population grew, the Spanish-language media grew, and I also grew as a newswoman.

In many ways I had no choice. Early in my career I had seen this population's predicament, its lack of political power. I realized that we in the Spanish-language media carried an enormous responsibility to reach this audience, not just for the sake of our ratings but for the sake of their survival. Every newsroom in the United States has its challenges. But when you have millions of viewers who were born on foreign soil and speak a different language from the American mainstream, your challenges are quite different. Hispanic viewers in the United States have specific issues they want to see explored on their screens. They want a mix of news, headlines from their homelands, but also vital information about their newly adopted country. They want to understand policy changes and how they will affect their families, immigration laws, social services, health and educational systems. They want to know what their rights are and how they can participate, socially and politically, in their communities.

Granted, there's a large part of our population born and raised American. And there are immigrants who achieve great success as professionals and leaders in business, law, medicine, the arts, sports, education, what have you. But our core audience is poor and largely uneducated. They need to know the basics, how to get their kids into school, how to deal with hospitals, how to access whatever resources are there for them. Many of them are eligible for services they have no clue even exist.

When I first started at KMEX, Latinos made up about 25 percent of Los Angeles's population, yet we had no clout or political representation. There were no Hispanics on any of the important local boards—the City Council, the Board of Supervisors, the Board of Education. And just as Latinos were reaching majority numbers in key areas of the city, the City Council dealt us a crippling blow in September 1982. By unanimous vote, the council approved a redistricting plan that diffused Latino strongholds, most notably in the 14th District, which covered significant parts of the "Eastside," East Los Angeles. The council carved out Hispanic areas from the northern part of the 14th and tossed them into the district that represents the San Fernando Valley. The 14th District also lost a vital voting block when the city sliced out a Latino public housing project in the southern end and simply gave it to another district.

So it was little wonder that a council veteran named Art Snyder, an old-school Irish politico, was able to hold on to the Eastside district in spite of its growing Hispanidad. With his unmistakable redheaded looks, Snyder was a familiar and well-liked presence on the streets of East L.A.—"El Colorado," we called him. And he was determined to stay. He clung to his seat even as a young urban planner from El Sereno, Steve Rodríguez, made a run for it in the 1983 district elections. Rodríguez, who had made headlines in 1979 when President Jimmy Carter stayed at his home during a visit to Los Angeles, gave Snyder a run for his money. He was four votes shy of forcing Snyder into a runoff. If Rodríguez had won, he would have become Los Angeles's first Hispanic city councilman in twenty-one years. But, thanks to the magic of redistricting, he didn't.

Art Snyder remained in his seat until he resigned for personal reasons two years later. By that time, the 14th District boasted some 200,000 residents, three-fourths of them Hispanic. However, barely 30,000 of them were registered to vote. The greatest voting block remained in the Anglo stronghold of Eagle Rock.

When Snyder's seat became open in 1985, Rodríguez ran again. This time his competition would be more formidable than an old Irish politician. The bigger threat would come from a fellow Latino, a tough-talking longtime state legislator named Richard Alatorre. Rodríguez and Alatorre couldn't have been more different. Rodríguez was a so-called Mexican Yuppie, while Alatorre, in tight with the political establishment, was a rough-and-tumble sort who dressed in fine Italian threads. In the end, he delivered Rodríguez a resounding defeat and became the first Latino on the City Council in twenty-three years.

In an ironic twist, just two weeks before the election, the U.S. Department of Justice had slammed the city with a lawsuit alleging "a history of official discrimination" against Hispanics. The feds took specific aim at Los Angeles's redistricting plan, accusing the city of gerrymandering districts to intentionally disperse the Latino votership and splinter its political power.

In response, the newly elected Alatorre was appointed chairman of the City Council's charter and elections committee, which was to review the controversial reapportionment plan. Districts were drawn and redrawn, and debates over them would churn for years after that as the population grew and, ever so slowly, more Latino names began to pop up on the council dais.

On the day of the district elections of 1983, I was sent out to do a man-on-the-street kind of story, collecting voter reactions to the prospect of having a Latino in City Hall. I walked up and down Lincoln Heights with my photographer, asking the same questions over and over:

"Are you voting today?"... "What do you think of the candidates?"..."Who are you voting for?"

I interviewed fifteen people before I found one who even knew there was an election going on that day. Most of them had no clue, much less a voter registration card.

"What election?" they'd say. I wanted to put my microphone down and talk some sense into them.

What do you mean "What election?" This is THE ELECTION. Why aren't you registered? I understand there were undocumented immigrants who could not vote, but what was everybody else's excuse? More urgently, how was I going to pull together a story with just one sound bite?

I huffed back to the newsroom and found Pete Moraga.

"I can't do this story," I told him. "Nobody knows about this election. Nobody cares. Nobody is participating. It's not a story."

Pete gave me a look I'll never forget.

"That is the story. It's there, right in front of your eyes. How can there be a Hispanic in City Hall when his constituents don't even know there's an election? That's your story."

And, indeed, that was the story. Right then I realized

there was an enormous gulf between this population and the systems of government. Yes, this population was alienated and disenfranchised because it lacked representation. But if it lacked information, that was our fault. It was our responsibility as Latino journalists serving these viewers, readers, listeners, to deliver the information that is vital to their lives. In this way, we could help connect them to the process.

This is a belief that has guided my career. Some may criticize it and argue that as journalists we are not supposed to be active in political campaigns. I'm fully aware of that. I don't believe in telling people who to vote for, but I do believe in telling them they should vote and giving them what they need to make an informed decision.

Following that 1983 election, I addressed my duties as host of *Los Angeles Ahora*, the public-affairs show, with added determination. Little by little, the feedback from the street told me we were doing something right. It felt good when somebody came up and said, "I got this job because I saw it on your program. Thank you." Undoubtedly, we were an important community liaison.

I admit that in the beginning I found it odd when people on the Eastside yelled out my name. I remember once a couple wanted to take a picture of their little girl with me. I just nervously scurried away, embarrassed. I thought to myself, *I'm not an actress or a singer or an entertainer. So why are people ask-*

ing for my autograph? Then somebody told me that if I didn't pose and sign autographs, I'd come across as arrogant.

"You have to respond to these people. Remember that they watch you on television. They see you differently than you see yourself."

So I relented. I was no less embarrassed, though, especially when the other reporters fired looks my way. Regardless, I knew that my role as a reporter, and later on as an anchor, would be different from that of my English-media colleagues. Not only did we in the Spanish-language media have a responsibility to cover the news, we also had to reach out to an entire population of people who felt disconnected from mainstream American society. Throughout the years, it's a responsibility I have taken very seriously. More than two decades would pass before I could witness the kind of political representation my early viewers had dreamed of. But I did, and proudly so, the day Antonio Villaraigosa became Los Angeles' first Latino mayor since 1872. The summer day he took his oath of office, he was joined by four other Hispanic city officials—the city attorney, the city clerk, and two councilmen. It was a day I could barely imagine in the early 1980s as I reported from the ranks of the city's invisible population.

IT TOOK me a while to hit my stride as a news reporter and anchor. But eventually, about four years into my TV career, I did. And just as I did, my world collapsed. My father got sick. His condition deteriorated quickly as he suffered from an illness I had never heard of, something called vasculitis, an inflammatory disease of the blood vessels.

As his health worsened, I tried to remain at his side as often as I could. I would split my time between the station and my parents' house. My mother and I braced for the worst as we watched my father gradually lose his ability to walk. He could no longer go to his beloved L.A. Dodgers games. Instead, he would listen to them on the radio.

For the last six weeks before his death in 1985, my father spent his days nearly motionless in a hospital bed, tethered to life-support machines. Doctors gave him a 1-percent chance of survival, but we clung to that 1 percent. One of my coworkers was a Christian preacher and I asked him each day to pray. "You're not praying hard enough!" I'd tell him whenever the news grew worse.

He tried to explain to me that even some prayers must be realistic—sometimes you have to pray for rest and a smooth passage to heaven. But I didn't want to hear that. Many years later, I would come to question the decision to keep my father on life support for so long. Were we extending his life, or prolonging his agony? At the time, however, I couldn't bear the thought of losing him without a good fight. When the doctors told us there was a chance of prolonging his life by inserting a tracheal tube, we approved the procedure. But it was a decision I would come to regret. In his final days, I watched this man of God grow so desperate that he scribbled a cryptic message on a scrap of paper. "The devil!" he wrote in an uncharacteristically shaky hand, pointing to the tube in his throat. That image has haunted me for years. When the late Pope John Paul II was fitted with a similar device, I couldn't help but remember my father's hellish days before he passed on to a better life.

And even when my father finally closed his eyes and rested, he remained with me in my thoughts and inner struggles. As my career took me to new places and experiences, I continuously wondered what my father, the perennial student of history, might say about my covering a presidential summit, a natural disaster, or a papal visit. I strained for answers that rarely came.

Send me a sign, Papi.

FOUR

❦

Birth of a Network

KMEX was a tiny station, but it was the flagship operation of the Spanish International Network (SIN). "Network," of course, was a relative term because there were more missing links than there were actual links in the SIN chain. The actual links were San Antonio, Miami, New York, and Los Angeles, all established by a handful of visionaries. While people in the mainstream dismissed us as irrelevant, those early players of Spanish-language television sized up their audiences and, in them, saw the future of the nation. Theirs was a rapidly growing viewership that was fiercely loyal to the low-budget, folksy shows that played on their TV screens each night. This viewership couldn't care less about ratings, glossy programming, or the established measures of

TV success. They cared about trust, and they could trust the familiar faces and personalities they saw on their favorite shows, most of which were imported from Mexico. In the early days of SIN, the U.S. stations shared a rather incestuous relationship with the czar of Mexican television, Televisa's then-chief, the late Emilio Azcárraga Milmo. As a result of business ties, convenience, and a lack of funds for original programming, they ran Televisa's international news, telenovelas, and entertainment shows. That relationship proved to be troublesome, as FCC laws prohibited foreign ownership of U.S. broadcast stations. In their dance of complicity, these stations and their executives were deemed to be little more than *"prestanombres,"* fronts for Azcárraga, as a lawsuit uncovered in the early to mid-1980s. How Univision emerged from this scenario is a long and winding story, a book of its own. But the condensed version is that Azcárraga and his proxies were forced to put the U.S. station group on the market in 1986. The prospective buyer was Hallmark, the greeting-card company from Kansas City, Missouri, which agreed to continue to buy SIN programming.

By then, I had made quite a few appearances on the network, either through stories, specials, or anchor fill-in duty for Teresa Rodríguez, co-anchor of the national newscast, which had been moved from Washington, D.C., to Miami. Safely removed from the power struggles and legal drama transforming the network's top echelons, I found the diversity within the stations of SIN and early Univision to be fascinating. So many distinct worlds—the Puerto Ricans and Dominicans in New York, the Cubans and Nicaraguans in Miami, the Salvadorans in Houston and Washington, D.C.—all remained glued to the same soaps, contributed to the same charitable telethons, and shimmied their *"colitas"* to the campy tunes on *Sábado Gigante*.

These cultures thrived beneath the radar of the official U.S. census predictions, the demographic studies, and the spin of the mainstream punditry. Meanwhile, their trusted TV personalities had also yet to register on the radar of the larger broadcast industry. But we, those early reporters and anchors of Spanish-language television, saw it coming, this tidal wave that would transform the United States. And there was no place that I wanted more to be than riding the crest of this wave.

However, in 1986, before the Hallmark deal was struck, a culture clash rocked Miami. Network news staffers, long resentful of Televisa influence, caught wind of a disturbing rumor: Televisa chiefs were planning to send Mexican anchorman Jacobo Zabludovsky to Miami to take over the network newscast. The idea of answering to Zabludovsky, who in their eyes was a mouthpiece for Televisa and by extension the Mexican government, triggered a rebellion. Under the leadership of veteran newsman Gustavo Godoy, the network news staff took pride in its aggressive reporting and objectivity, alien concepts at that time in Mexico, where the government controlled news content. The Zabludovsky scenario most likely would mean the ouster of Godoy and other high-level personnel, not to mention the fact that it would compromise the credibility of the newscast. So the Miami staffers fought back, Miami-style. They took to the airwaves on local Spanish-language radio and stirred up the Cuban exile population with this alarming notion: *Word is Jacobo Zabludovsky, mouthpiece of the Mexican government, is coming to Miami to take over our newscast.*

While this newsflash may seem mild on its face, consider the fact that a great many people in the Cuban exile community detested the Mexican government for its cozy ties to the Fidel Castro regime in Cuba. Mexican telenovelas and variety shows were one thing. But Mexican news spin was quite another. Miami's fury, picked up by the national press in New York and Los Angeles, spread across the country and into other SIN newsrooms. On the West Coast, where few cared about the Castro government and its allies, opposition to Zabludovsky was equally strong, although for other reasons. We didn't trust the Mexican media in general because of its progovernment bias.

The Miami revolt hit the network hard. A large group of

staffers, led by Godoy, left SIN and launched a competing network, the Hispanic Broadcasting Corporation (HBC). That split would prove to be an earth-shattering incident in my career. I was still anchoring at Channel 34, filling in at the network only occasionally. My KMEX colleague, Jorge Ramos, was hired away by the network to cohost a morning show called *Mundo Latino*. I had been told that the network wanted me to be a part of that show, which would be presented from Miami, Los Angeles, and Mexico City. But my general manager, Danny Villanueva, nixed the idea. I had been at the local station for six years and I wanted nothing more than to move up to the network. I was furious at Villanueva. I was on my way to his office to resign when I bumped into him on the stairs. When he saw the look on my face, he threw his arm around my shoulders.

"So I heard they told you I vetoed the *Mundo Latino* idea," he began. "You know what? You belong in news. You're a newswoman. I don't want you to do that fluff show."

Danny Villanueva was a tough-talking jock, a former pro football player who got his start in TV as a sportscaster and worked his way up to the top job. He was also, unwittingly, a bit of a psychic.

"That show isn't going to last, anyway," he declared. "It'll be canceled, and then where will you be?"

It turned out he was right, though of course neither

one of us knew it at the time. But I didn't want to hear it, and his pep talk didn't make a dent in my resolve to break into the network. Yes, I was a newswoman, but I wanted to grow.

Meanwhile, Jaime Dávila, a high-level Televisa executive and Azcárraga's troubleshooter, tried to salvage what was left of SIN by offering handsome contracts to those who had stayed. Network heads feared the newly formed competition would drain them of their remaining talent. And they were correct in their suspicions. In December 1986, not too long after Gus Godoy had assembled his troops at a sprawling warehouse in Hialeah, he flew to Los Angeles and reached out to me. An affable guy admired by his reporters and anchors, he made me an offer I couldn't believe: a seat on the anchor desk. I would be cohosting the evening newscast with a good-looking Uruguayan commentator named Jorge Gestoso.

"He's the Latin Peter Jennings," Gus told me. Not a bad thing to say to a Jennings fan.

Gus made this sound like my dream job, but there was a problem. I had never lived anywhere but Los Angeles, except for those childhood years in Mexico. My whole life was out west—my family, my friends, my roots. What was there in Miami for me? Gus wanted to talk about it further, and I was happy to listen. I felt honored that he believed in me. So I flew to Miami, where he introduced me to one of the owners of the new network, a financial officer who looked no older than eighteen. If Gus was the welcoming and generous news director who offered me the moon and the stars, Charlie Fernández was the humorless bean counter who brought me back down to earth.

"I don't know why Gus offered you all these things," he snapped. "I don't see why we should do this for you."

By "all these things" he meant the safety clause my attorney was asking for in the contract negotiations. We asked that HBC give me a three-year contract guarantee, in the event that the network went bust. I thought it was a reasonable request. But the financial guy begged to differ. He treated me as if his company was doing me a favor in even considering me for the job.

Nevertheless, after much soul searching, I accepted the anchor position. When I returned to my station in Los Angeles, I went in to see Danny Villanueva and resigned.

"I've decided to go work with Gustavo Godoy at HBC," I told him. Villanueva played it cool.

"Okay," he said, "but just one thing before you leave..."

He told me Azcárraga's troubleshooter, Dávila, wanted to talk to me. So Villanueva called Dávila, handed me the phone, and walked out of the office.

Dávila quickly got to the point. He wanted to hire me to

anchor the network's new late-night newscast. It was an interesting offer for two reasons: it showed me the network brass wanted me to join their network and it also revealed their desire to detach from the whole Zabludovsky affair. The new show would replace Televisa's late-night newscast in the United States. That newscast, *24 Horas*, anchored at the time by Mexican newswoman Lolita Ayala, had once been the domain of Zabludovsky.

I politely thanked Dávila for the offer but declined. I wondered why they had waited until after I resigned to make it. He asked that I give him a couple of days. He was coming to L.A. and wanted me to meet someone. Fair enough, I thought. I called Godoy to tell him SIN wanted to talk and that I thought I owed it to them to listen and to myself to explore all my options.

"Go ahead. I understand perfectly well," he said. But he gave me a deadline.

The following Monday, I met with SIN's Dávila at a Beverly Hills hotel suite. After we chatted for a little while, he excused himself and went into the other room. He came back with a tall, distinguished gentleman.

"Meet the new head of the network," Dávila said, introducing the other man. He was Luis Nogales, a U.S.-educated Mexican-American they had brought in from a top position at United Press International. And he had big news: Zabludovsky was out of the picture. His former show, *24 Horas*, would be canceled.

"We are no longer importing journalists from Mexico," Dávila told me. "We want homegrown journalists from the United States, people such as yourself. You have a perspective that Mexicans don't have because you were raised here."

They said they were bringing in a new news director, a former UPI foreign editor named Sylvana Foa. It was a brand-new day at SIN. Finally, the ripple effects of that initial Miami pressure had reached the top of the network. And there, in a swank corner of Beverly Hills, stood its future. This was the point of demarcation, and the new leadership was inviting me to step into a new republic, one that would be defined not by Mexicans in Mexico, but by Hispanic-Americans in the United States. Televisa had realized that it could export their telenovelas, their entertainment shows, and their movies, but there was one thing it could not export to the United States: its news.

The rejection from his people in the States stung Azcárraga. He was a man accustomed to acceptance wherever he went. It's no wonder they called him "El Tigre." If he didn't have access, he could buy it, for he was one of the wealthiest men in all of Latin America. There was very little he couldn't have. And this was it—he couldn't buy the respect and acceptance of his news department from a diverse audience.

The SIN salary offer was good. It was certainly much more than I was making at KMEX. But when I told Dávila that it wasn't as much as the HBC offer, he didn't hesitate.

"We'll match it," he said.

And here was the biggest plus: I didn't have to leave Los Angeles, my home. That's what sold me on SIN. My lawyer, Jim Blancarte, had a different take. He didn't think it was fair to turn down HBC for the same pay. So he asked SIN to top the HBC offer.

As the network brass deliberated, I grew anxious. The deal was not yet resolved when it came time for the SIN Christmas party. As people danced and sipped their spirits, I kept calling my answering machine for messages. There were three from Gus in Miami.

"María Elena, I'm waiting for your reply."

"María Elena, I'll give you another hour."

"María Elena, if you don't answer me the deal is off."

But the SIN people had not yet gotten back to us on the salary issue. I found my lawyer at the party, and he hadn't heard anything, either.

I went home, still anxious. And my phone rang. I couldn't pick it up. I could only stare at it. I let the answering machine pick up. In my frozen state, I barely recognized the voice of my lawyer:

"María Elena, we have a deal. They topped the HBC offer."

I was relieved, but I wasn't jumping for joy. I felt horrible for Gus. I called him and gave him the news. I thought he would be furious, but instead he had gracious words.

"Thank you for considering my offer. My doors will always be open to you."

So I stayed, and we launched our new program in January 1987. Under a brand new name, Univision, the network put on a big press conference to announce that a homegrown Mexican-American, and not a Mexican import, would be anchoring a new late-night newscast. I had expected my job description to include a lot of things, but PR weapon wasn't one of them.

Azcárraga didn't own the stations, but he did own the network, and he wanted to protect his interests. He still had staffers in place from the thwarted Zabludovsky plan of months earlier. So he decided to move them and the entire newscast to California. He wanted nothing to do with Miami—for the time being. (Ironically, Azcárraga would not only come to love Miami but would die there, on is yacht in 1996.) His staff of Televisa executives and consultants was tossed in with the new people brought in by Nogales. It was culture clash, to say the least, pitting two cultures and two very different styles of reporting the news against each other. Add megadoses of three-o'clock caffeine, and it was a recipe for chaos.

Sylvana Foa, our newly appointed news director, fresh from UPI, might have known a lot about world events, but she knew nothing about television. She couldn't understand why we needed three cameras in a studio and two editing machines. The disconnect between her and the reporting staff was enormous. Worse still was the power struggle between her and the Televisa news executives. They would come in and tell us to do things one way, then Sylvana would come in and tell us to do things another way. The rivalry was intense, and we were in the middle of it. We didn't know who to trust or who to listen to.

Meanwhile, I remained friends with those SIN defectors at HBC. I was happy to see them in Los Angeles during the National Association of Hispanic Journalists conference that year. When I heard Gus was in town, I invited him to dinner. Big mistake. When we arrived at the restaurant, a Mexican spot on Melrose, we ran into some unexpected faces—my bosses at Univision. The next morning, I got called into Nogales's office for a verbal beating. He demanded to know why I would dine so publicly with the enemy. It was an act of betrayal, he charged.

"You insulted and degraded the company!" he snapped.

His harsh words brought me to tears. I tried to explain that Gus was a friend, someone I felt I owed the courtesy of my company and time. It didn't make a difference to Nogales. He couldn't see beyond the fierce competition between the networks.

While I understood the company's anger at the defectors, I didn't understand Nogales's overreaction to an innocent social situation.

"You are not allowed to talk to them, understand?" he warned.

The next day at the NAHJ conference, I hesitated when I saw my good friend Carlos Botifoll, one of the defectors. When he came over to say hello, I shot him a look that said, *Not now.* Carlos was so hurt he stopped speaking to me for a long while.

Back in the newsroom, the tension was unbearable. During a trip to the East Coast, I made a stop in Miami, hoping to catch Joaquín Blaya, general manager of Channel 23 and a key player at Univision, for an off-the-record chat. His schedule was so tight, I took what I could get, a few minutes in a VIP booth overlooking the Grand Prix of Miami.

"Mr. Blaya, our news department is falling apart. Somebody's got to do something," I told him.

None of it was news to him. He had heard about the power struggles. He told me to stay calm, that things would change. And they certainly did. Shortly thereafter, Nogales was out; Blaya was in. The dueling news executives were gone and in came Miami journalist Guillermo Martínez, a Cuban-

born veteran newspaperman who had worked as news director for Univision's Channel 23. Although he was relatively
new to the broadcast industry, he had surrounded himself
with a talented group of producers and had quickly soaked up
the rules and the lingo.

I had known Guillermo for many years, from the days
when we both served on the board of the National Association of Hispanic Journalists. He was president, I was vice-
president. I was hoping he didn't remember all the times I
voted against his proposals. But I had nothing to worry about.
He was grateful that there was someone on staff he knew and
could rely on. By this time our news department had moved
from Hollywood to Laguna Niguel, in Orange County. I became Guillermo's confidante, informant, and, for a while, his
chauffeur to and from work—and it was a long commute.

Guillermo quickly changed things around. He stream-
lined our staff and cut our one-hour newscast in half. And in
early 1988, he did something that forever changed my career:
he gave me a partner. He teamed me up with Jorge Ramos,
who had been co-anchoring the early newscast with a young
Uruguayan journalist named Andrea Kutyas. Per Guillermo's
new order, Jorge and I cohosted both shows, the 6:30 P.M. and
the 11:00 P.M. Jorge and I had been working together since
1984. But this is how we came to share the anchor desk.

I felt honored, although I also felt a pang of indecision.

My good friend Teresa Rodríguez, who had been SIN/ Univision's first female anchor, had been on maternity leave in Miami, where her husband was an executive at the local affiliate. I wondered if she was planning to return to her former post. She assured me she had no intention of leaving Miami and that she had plans to work on periodic specials. With that, I jumped into the job.

The first thing Jorge and I did was establish the rules of the anchor desk. Over dinner one night, we made an agreement: we'll take turns. We'll take turns opening the show. We'll take turns conducting big interviews and reporting major stories. The only thing that would remain the same is our places at the desk—I would sit on the left, he would sit on the right. He would get the pink copy of our multicolored script and I'd get the yellow.

In network television it's rare enough to find a male-female anchor team. And in the macho world of Latin American television, women anchors are often considered merely ornamental. Sure, a woman might be permitted to read a few stories, but rarely would she be allowed to rise to any level of prominence. So for me it was crucial that Jorge and I be considered co-anchors in every sense.

When we returned to the office, we told Guillermo about our agreement. He was pleasantly shocked, pleased to see we weren't going to be a couple of bickering prima donnas. "You're the most civilized journalists I've ever met," Guillermo declared.

And to this day, seventeen-plus years later, Jorge and I have honored our vows. It certainly helps the situation that although Jorge has developed into a successful author and respected opinion leader, he remains a gentleman and a mindful coworker.

Three years later, in a move to centralize its operations, Univision moved the newscast back to Miami and I was forced to leave my city behind. To say that I didn't want to go is an understatement. There was a nasty buzz around the network, insiders warning us to be careful with the Miami Cubans. Steer clear of them, it went. Don't listen to Cuban exile radio. Don't live in Dade County, if you can help it. I didn't know what to expect, but I had a hunch these were nothing more than alarmist exaggerations. Finally, it was Joaquín Blaya who confirmed my suspicions.

"Please. You'll love it here," he assured. "You'll find yourself a nice Cuban guy and get married."

How could he possibly have known?

FIVE

❧

Juan Diego's Miracle

I'm not an overly religious woman, despite my Catholic upbringing. I have faith, yes, but I only go to church on Sun-
days occasionally, and like many modern-day Catholics, I take
issue with some matters of skewed dogma. For example, after
sitting through a thinly veiled political discourse by a visiting
priest who condemned "abortionist" candidates one Sunday
just days before the 2004 presidential election, I felt so disconnected from the Church that I questioned if I would ever
return. Especially after the third time the priest thundered to
the congregation:

"We have to vote out the baby-killers!"

I sank in my pew. For one thing, I believe politics has no
place at the pulpit. But, more important, I didn't want to have

to explain abortion to my seven-year-old daughter, whose catechism assignment that week was to take notes on the Gospel and the sermon. We haven't even discussed "the birds and the bees."

But, no matter my level of discontent, I can never get too far away from the Church. I have covered at least a dozen papal visits and hosted two one-hour specials about the Catholic Church. I pray with my daughters every night, and when I travel, I make the sign of the cross before my flight takes off. I always carry my Virgencita de Guadalupe. I keep her image at home, in the office. I take her on the road. I light a candle to her every December 12th, her feast day. It's not necessarily the Catholic schoolgirl in me that's so devoted to the patroness of the Americas, it's the Mexican in me. More than a religious symbol I see her as a cultural icon, a likeness of Mexico itself, lovely and mestiza, a Madonna rising in an oval of glorious hues, rays of the sun, and roses. She was my mother's shoulder to cry on, and then she became mine, watching over me in difficult times. And the story of her favorite son, Juan Diego, the poor Mexican Indian to whom she appeared in 1531, holds a particular appeal for me. Like all the apparitions of the Blessed Mother, this one is a poetic story.

Juan Diego was an indigenous man from a small village just north of Mexico City. The story, as recorded in sixteenth-century Aztec literary documents, is that this humble farmer and weaver lived during the Spanish conquest of 1521 and, later, swayed by early Franciscan settlers, converted to Christianity. At his baptism, he shed his Aztec name and took on a new one, Juan Diego. One wintry morning as he walked to church, he heard a hypnotic voice coming from a nearby hill. Amid the chirping of birds a woman's voice called his name. He followed the voice up the slope of the hill, and when he reached the top he found a woman of staggering beauty, an Aztec princess who spoke to him in his native language. This ethereal princess was the Blessed Mother, and she asked Juan Diego to relay a special message to the bishop of Mexico: she wanted a shrine built on the very spot where she stood.

When Juan Diego delivered the message, the bishop didn't believe him. He asked for a sign. Days later, the Virgin stopped Juan Diego once more at the foot of the hill. She instructed him to climb the hill and gather flowers for the bishop. When the farmer reached the hilltop, he was astonished to find bushes of rare roses. He picked the flowers and bundled them into his *tilma*, the rough-hewn Indian cloak he wore, and took them to the bishop. As Juan Diego unfurled his cloak, the roses tumbled to the floor and, miraculously, the image of Our Lady appeared emblazoned on the front of his *tilma*. That image of the Virgin dressed as an Aztec princess sent a strong message to the Spanish conquerors: their triumph depended on the oppressed indigenous people of Mexico.

In the years following the apparition, millions of Indians, moved by Juan Diego's story, converted to Catholicism. But although the story was embraced by the Catholic Church throughout the ages, many doubted the Aztec farmer ever existed. Chief among the skeptics was none other than Father Guillermo Schulenburg. Not only was he a Catholic priest, but he was also the abbot of the Basilica of Our Lady of Guadalupe. He suggested that the Juan Diego story might be a myth. Those claims prompted the Vatican to assemble an investigative commission. In 1998, that commission reported its findings to the Vatican Congregation for the Causes of Saints. It's conclusion, backed by historical Indian documents: the Juan Diego story was true.

That controversy was still years away when my crew and I traveled to Mexico City in May 1990, braving the mobs and media frenzy, to cover Juan Diego's beatification by the late Pope John Paul II. And it seemed fitting that we would need a little help from the blessed farmer himself. No, getting our press credentials was no problem and neither was securing the perfect spot on the roof of a Mexican restaurant overlooking the Basilica of Our Lady of Guadalupe and, just behind it, the hill of Tepeyac, where the Virgin appeared to Juan Diego.

We needed her humble servant to get the pope to stick to his schedule.

You see, we had put together a one-hour special to coincide with the pope's arrival. It was big news for our viewers in the United States, where nearly 8 million Mexicans lived at that time. This was the pope's second trip to Mexico, home to the world's second-largest Catholic population—Brazil has the largest. And, judging from the overwhelmingly warm welcome during his first trip, this visit promised to be one for the history books. Not to mention that Juan Diego was well on his way to becoming the first full-blooded Mexican Indian to reach sainthood.

Getting our special on the air would require some meticulous planning, maneuvering, and, yes, Juan Diego's help. The pope was scheduled to arrive at noon. His plane would land on Mexican soil a few minutes earlier and, once diplomatic protocols were completed, he would descend from the aircraft. Our special broadcast would begin at 11:30 A.M., airing interviews and reports about the extensive preparations leading to His Holiness's visit. But we ran into a bit of a problem. The Mexican government decided it would take control of the airwaves for the occasion and transmit the pope's arrival ceremony on national television, and it would strictly limit access to all foreign media. So the only way we could get our show on the air was to pretape the first half-hour. We were monitoring the live pictures from the Mexican government's news-pool feed for the first sign of the pope descending from his airplane. That would be our cue to cut to those live pictures, over which we would provide live audio commentary.

We covered all ground. Or at least we thought we had. I had taped a brief introduction as a lead-in to those first live shots of the pontiff. If he happened to exit the plane a few minutes earlier than scheduled, we would cut to the tape and there I'd be with my intro. I'd turn to the camera and reverently announce:

"*Y ahora llega a suelo mexicano el Papa Juan Pablo II.*"

Of course we were relying on the pope's infallibility. He is never late. And so our show began. For twenty-five minutes, our viewers saw a mix of stories on Juan Diego and the beatification process. There were testimonials from people who had walked all night from distant villages just to get a glimpse of the pope, in the "pope-mobile," winding through the streets of the Mexican capital en route to the papal nuncio's residence, where John Paul II was to spend the night. As the tape rolled, we stood by in a studio audio booth, watching the monitors for our cue. At my side was a young priest who had close ties to church hierarchy. He was ready to help narrate the arrival of the Holy Father and identify faces in his entourage and in his welcome committee.

As high noon approached and pool images streamed by on the monitors, we realized the pope's plane, which had landed on time, was still just sitting on the tarmac. It wasn't moving. The door wasn't opening. The pope wasn't descending. And our taped material was running out. In just a few seconds, the tape would reach the part where I'd announce: "And now we go to His Holiness, Pope John Paul II, arriving on Mexican soil."

That's when my producer, Marilyn Strauss, began to pray to Juan Diego.

"We need a miracle," she beseeched.

As our tape spun toward its end, I cringed. Marilyn, her eyes fixated on the image of the motionless plane, held her breath. I was nervous, but I thought she was going to have a heart attack. After all, we'd look like utter fools if we drumrolled to the pope and there was no pope. And sure enough, although the aircraft's door hadn't budged, my intro cued up and there I was before millions of viewers worldwide announcing:

"Y ahora llega a suelo mexicano el Papa Juan Pablo II."

At that very instant, as if following a more authoritative cue from the heavens above, the door of his plane swung open, framing the figure of Pope John Paul II. Marilyn gasped in disbelief.

"It's Juan Diego's miracle," she cried, and we exhaled together.

Our special coverage of the pope's arrival was a success. But, unbeknownst to me, the real surprise was just around the corner. As we made our way back to our Mexico City bureau, I struck up a conversation with the young priest who had been my cohost that day. His expertise made our coverage more complete, without a doubt. I thanked him for his help. Then our conversation took a more personal turn. I was no longer the journalist discussing coverage, I was the confused Catholic sharing my doubts and questions. I told him about the documents I had read after my father's death, the ones in his Box of Secrets. For some reason, after my discovery I couldn't resist sharing the story with just about every cleric I encountered. Call it an informal confession of sorts, but each time I interviewed a Catholic priest for one story or another, I felt compelled to tell him two things: the fact that my father had been a priest in his youth, and the fact that I was divorced. I had a serious issue with the Catholic Church's intolerant stance on divorce. I felt it was unfair that the Church should censure a practicing Catholic simply because his or her marriage failed.

"I'm sure God would not want me to be in an unhappy marriage," I'd tell the priests.

And as for my father's story, I was haunted by the words my mother told me when I confronted her with the documents—"He suffered a great disappointment that led him to leave the Church," she had said.

What could have possibly happened? I always wondered.

I questioned my motives in wanting to know the truth. What was driving me, a daughter's confusion or a journalist's curiosity? After much soul-searching, I realized that I just wanted to know. Period. Whatever it was that led my father to leave the priesthood didn't matter. I would not judge him for it. But I needed to know what happened. My identity depended on it.

After hearing my story, the young priest, a member of the conservative Opus Dei movement, offered to help. "Maybe I can find out some information about your father," he told me as we said goodbye that day.

A couple of days later he called to say he had something to tell me. He asked me to come down to see him at the rectory of his church. When I got there, he told me he had done some research on my father and had located a parish where he had once served. It was a small church on the outskirts of Mexico City.

His words jolted me. Even though my mother had confirmed my suspicions about my father's priestly past years earlier, this was the first time I had some kind of independent confirmation. I tried to imagine my father dressed in a priest's cassock, walking among his faithful. But I couldn't. I couldn't reconcile my memories of him, even those in which he strolled in the park, lost in prayerful meditation, Bible in hand. I just couldn't see it.

My church source told me there was an elderly priest at my father's former parish and that he'd probably remember him. But then he turned somber and gave me a warning.

"You should let this go," he said. "If he didn't tell you, it was for a reason and you must respect that."

I asked him for the name of my father's former parish, but he declined to give it to me.

"Let your father take his secret to his grave," he said.

Then why did he call me over to the rectory in the first place if he wasn't going to share his findings with me? Why would he put me through such torture, especially after I told him how important it was for me to know more about my father's past?

"What I really want to talk to you about is your divorce: you must find a way to get back together with your husband." I couldn't believe what I was hearing.

"But my ex-husband is already married to someone else," I told him.

It didn't matter. The priest gave me a stern lecture on commitment.

"When you marry someone, you marry them for life," he said.

I thanked him for his time and left the rectory, thinking, What could this guy possibly know about failed marriages? I was frustrated and fully convinced he had wasted my time.

But in fact he hadn't. In the years to follow, the sliver of information he gave me, that my father had been stationed at a parish outside the capital, would rekindle my hopes that I could uncover the truth about his decision to leave the priesthood. And it would nudge me along in my search. For the time being, however, I was stuck on that second part of his message. Reconcile with my ex? Absurd.

I HAD divorced at twenty-six and written off marriage. As my job began to consume more of my time, I rarely gave the notion a second thought. The idea of motherhood, however, was never too far from my mind. It was my most cherished wish, and had been for as long as I could remember. I was fourteen years old when my sisters had their babies, and even then the idea tugged at my heart. Of course, I had vowed not to stumble into motherhood as an inexperienced teenager, as they had done. In my first marriage, my husband and I had decided we would wait two years before starting a family. Our marriage lasted a year and a half.

But as my single days stretched into eleven years, my hopes of becoming a mother dimmed. And then came Eliott Rodríguez. He was a handsome, quiet guy who worked as an anchor and reporter for WPLG, Miami's ABC affiliate. We were introduced by Marilyn Strauss, who has been my field producer for many of my most important stories and is also a close friend. This time we were all on assignment in Madrid, covering the July 1992 Ibero-American Summit.

"That good-looking guy from Channel 10 is coming," she hummed one night before dinner.

"Yeah, but he's married," I told her.

"Not anymore," she said.

Marilyn was a suave, crafty Cupid. She invited him to ride with us to interviews and join us for dinner. She'd grab the seat next to him and save it until I got there, then casually scoot over so I would sit next to him.

He didn't pay much attention to me at first. It wasn't until after we got back to Miami that he called and asked me to go out. On the night of our first date, he told me his divorce had become final that day. He brought pictures of our trip to Spain and, as we leafed through them, we got to know each other a little better. Soon enough, we clicked. I realized we had a lot of things in common. We loved our families and we loved our jobs. He was the son of hardworking Cuban immigrants and I was the daughter of hardworking Mexican immigrants. He was raised in the South Bronx, I was raised in South Central Los Angeles. And we had entered the news business at about the same time. He was down-to-earth and charming in a quiet way. He had two daughters and spoke glowingly of them. He was a wonderful father and I think it was this quality that won me over.

Shortly thereafter a group of us went down to Puerto Rico for the wedding of my co-anchor, Jorge Ramos to his beautiful bride, Lisa Bolívar. My friends, still playing Cupid, invited Eliott to come along. And there, between the palms on the crooked streets of Old San Juan, our romance blossomed. It was the sweltering height of August, and summer storms churned in the seas nearby. One in particular, an ambitious system named Andrew, seemed headed for Florida. On the night before the wedding, some of us went to a casino. We returned to our hotel rooms at about 2 A.M., only to find an urgent message from our news director, Guillermo Martínez, who was in town for the wedding, along with the top producers of our newscast. We would all have to take a morning flight out of San Juan, he said. Hurricane Andrew was speeding toward South Florida.

The next day, our entire newscast group—with the exception of the groom, of course—flew to San Antonio, where Univision had an uplink to broadcast our storm coverage. If Andrew was as powerful a storm as forecasters predicted, it could knock our Miami home base out of power. Poor Jorge. He had invited very few of his friends to the wedding, and now most of us were off to cover a hurricane.

This was one story Jorge would have to skip. As he bid us farewell outside the church, he excused himself: "I have a wedding to attend."

MEANWHILE, ELIOTT took the last flight to Miami, to check on his daughters and his home, and to cover the storm for Channel 10. Hunkered down in San Antonio as the first storm pictures flashed in, I was worried sick about my mother and my house, which was just one block away from Biscayne Bay. Just as I was getting ready to go on the air with a bulletin, I got a call from Miami. I learned my house had been virtually destroyed. Still worse, the news footage showed swaths of devastation across South Florida, homes battered, entire neighborhoods wrecked as if by a nuclear bomb. I could feel my face redden and my eyes begin to tear up. My mother was safe with other relatives, thank God. But my home—not just my house but my city—was no longer recognizable. I was afraid I was going to lose it on the air. Someone ran out and brought me some tissues.

A couple of days later, Eliott called to say he had checked on my house. He said the flooding had been so severe that my furniture floated into the street. I had a titanic clean-up effort ahead of me upon my return. The refrigerator in my guest cottage had floated up and plopped itself sideways, blocking the front door. As I struggled with it, my savior arrived in his finest work attire. I stood in amazement as he rolled up his sleeves, got down on the floor and pushed up the fridge. He turned it right side up and shoved it out of the way.

Wow, I thought as I watched him, a man who would do this is worth keeping around.

And so I did.

SIX

Angels and Babies

Puerto Vallarta is a jewel of a resort town, nestled between the foothills of the Sierra Madre and the rugged edge of Mexico's Bandera Bay. From the terrace of the Krystal Hotel, the Pacific Ocean appears between seven Roman pillars that rise from a poolside rotunda. It's a glorious place, and it is where Eliott and I decided we would exchange our vows on March 7, 1993. We chose Mexico in March to coincide with my mother's birthday, March 4th. She hadn't been back to her homeland in many years and, as she was weakened by a stroke and getting on in years, she wanted to visit with her brothers and sisters. It seemed like the perfect opportunity. We wouldn't have to worry about having a big wedding. We could have the family together and invite close friends.

In the few days preceding the wedding, some of the fifty

friends and relatives invited to the ceremony began to stream into the romantic city of cobblestone streets, whitewashed villas, and spectacular views. Eliott was to arrive later, as were my sisters, my best friend, Regina Córdova, my producer, Marilyn Strauss, my co-anchor, Jorge, and his wife, Lisa, as well as other friends from Los Angeles. My uncle Rodolfo, my mother's brother who lived in a town nearby, was to walk me down the aisle as a guitar trio played Mexican love songs. We had planned everything perfectly. Well, perhaps not everything.

I was thirty-eight years old, and I was three months pregnant. The pregnancy had been unexpected but certainly welcome. I admit I was a little scared at first. Eliott and I had only known each other for seven months. The reality seemed daunting. But after dreaming of having a baby for so long, motherhood was finally within reach. I would have had no problem with being a single mother, but Eliott and I decided that we wanted to be married. After a dozen years of being virtually married to my job, I had finally found a man I wanted to share my life with.

We had told only a handful of people, my mother and sisters among them, that I was pregnant. At first my mother had trouble accepting the news, but soon enough she gave me her blessing. She had known for years that this was what I wanted more than anything in life.

Three days before the wedding, my relatives and I planned to drive to the small town of Tepic in the state of Nayarit, where my uncles lived, to celebrate my mom's birthday. This was an arduous trip for her. It was the first time she had traveled since her stroke. After my father passed away, it took months for us to convince her that she needed to go on with her life. She had always been a lively, socially active woman. She loved to spend time with her friends, the same friends she had since her youth. And she loved to travel, mostly to Mexico to visit her brothers and sisters.

So this trip to Mexico was very special for her. She would get to see her few remaining siblings. She and my sister left for Tepic a day early, while I waited for Regina to arrive from California. But the night before we were to leave, I started getting cramps. The pain grew so intense that it caused me to double over in agony. And to make things worse, I began to bleed. When Regina saw me curled up on a corner of my bed, she called the hotel nurse. A doctor rushed over to see me in my room. Her diagnosis was devastating.

"You're having a miscarriage," she said. The only thought turning in my head was, *It's not possible. It's not possible.*

They took me to the hospital, where doctors wanted to perform a curettage. But I refused. I wanted a sonogram. I wanted to make sure that they were not mistaken. What if the bleeding was only a complication in the pregnancy? I wanted to believe that the baby I so longed for could be saved. I had to wait for a specialist to arrive the following morning. The sonogram only confirmed the bad news. I felt as if my heart had dropped out of my body and with it my sense of place and balance. They had given me morphine for the pain and I felt as if I was floating just slightly above the ground. I would look out the window and see this picturesque town of ancient streets and brilliant flowers, but it seemed to exist in some distant, mist-shrouded dimension. My wedding was two days away and I could do little more than lie in bed. Eliott arrived the following day, along with the rest of my relatives. But I told very few people what had happened. I couldn't even bear to tell my mother for fear that the news would impact her delicate health.

My wedding day was picture perfect. I walked down a path lined with palm fronds, escorted by my uncle. My stepdaughter Bianca, four years old at the time, led the way, sprinkling the aisle with rose petals. The ocean sparkled before us as we recited our vows. And just as the judge turned to Eliott and declared, "You may now kiss the bride," the sun sank into a swell of orange. It was beautiful. I'm sure my guests thought I was crying from sheer joy alone. But as a mariachi band played torrid *rancheras*, only Eliott and I knew how bittersweet that moment truly was.

THAT EXPERIENCE foreshadowed the emotional rollercoaster of the next five years, most of which I spent pregnant.

A few months after I lost the baby, I got pregnant again. This time I went to see Dr. Anthony Lai, who specialized in high-risk pregnancies. About eight or nine weeks into the pregnancy, I went in for a sonogram, and again I was met with alarming news. The baby wasn't developing properly. The doctor said I'd most probably miscarry again.

It's not possible. It's not possible, I thought once more.

I went for a second opinion, and that doctor came to the same conclusion. I was still not convinced. So I went to Dr. Neil Goodman for a third opinion. He's an endocrinologist who specializes in fertility. He confirmed that the fetus wasn't viable. Then came the wave of cramps and déjà vu. The miscarriage had come after a reporting trip, which led me to believe that traveling was what was causing me to lose my babies. I had flashbacks of all those bumpy roads in Puerto Vallarta—surely that's why I had lost that first child, I reasoned. Of course my biggest fear was that I had waited too long and my biological clock had run out its time. Maybe I'm just too old, I thought.

Nevertheless, from then on I became extremely cautious. Dr. Goodman put me through a series of tests to determine what was causing me to lose the babies. He prescribed prenatal vitamins, low-dosage aspirin to regulate my bloodflow, and monitored my ovulation cycles. He had me come to his office every month for a pregnancy test. I was determined to get pregnant and stay pregnant for nine months. Even as I prepared to embark on a reporting trip to Chiapas to cover the peace talks between the Mexican government and the Zapatista insurgents in March 1994, I took all kinds of precautions in packing my bags and scheduling my activities. I was supposed to go for my monthly pregnancy test on the Saturday before my trip. But on Friday night I began to bleed. I called my doctor and canceled my regular visit. What was the use, I told him, I can't possibly be pregnant, not if I'm bleeding. But the doctor insisted I come in for a blood test. Sure enough, he had good news and bad news.

"You're pregnant," he said. "And you can't travel." That's impossible, I thought. I had to go on this trip. This was a huge story. There hadn't been a war or revolution in Mexico in decades. I had never in my career said no to an assignment.

So I faced a dilemma: miss the story of a lifetime or risk my chances of having a successful pregnancy. Needless to say, if I knew traveling would mean losing the baby, I would have gladly given up the trip. But I talked it over with my doctor, who said I could travel as long as I took the proper precautions. I was not to carry anything heavy, not even a purse. I was to keep to a strict diet. And I was to have a daily shot of progesterone. I left his office with a bag full of syringes and the miracle hormone that might allow me to carry my child all the way through delivery.

I couldn't have asked for a better crew to accompany me.

Veteran cameraman Simon Erlich hovered protectively, carrying my luggage and smoothing my path. Producer Patsy Loris became my paramedic. With the help of a nurse in Mexico City, she learned to administer shots by practicing on an orange. The nurse told her to be careful, for even the smallest air bubble could kill me. Poor Patsy was more nervous about my daily shot than she was about getting the shots in Chiapas. Her hand shook every time she had to go through the ordeal. But she never complained. Even Porfirio Patiño, our Mexico City bureau chief, took care of me. He'd bring me fresh vegetable soup for dinner, and was constantly instructing the driver to slow down as we wove through uneven, mountainous roads on our way down from San Cristobal de las Casas to Ocosingo, where fierce battles had taken place.

I stayed away from the press conference, for fear of the usual shoving and elbowing of photographers and reporters. Besides, we had our correspondent, Bruno López, inside covering the breaking news. Some years later, I would have the opportunity to interview the Zapatistas' elusive leader, Subcomandante Marcos. But on that first trip to Chiapas I would remain on the sidelines, reporting sidebar stories and taking the pulse of the people. And what a story it was. I was so swept away by the historic undertow of it that I had no time to stress about morning sickness or any other surprises this pregnancy might have in store for me.

This was a pivotal moment in Mexican history, not only for the country's political life but also for its media. Up until that point, the Mexican media had been gripped by systematic censorship and bouts of self-censorship. Television stations, radio stations, and newspapers were virtual mouthpieces of the government. They frequently attempted to shake loose of the official grip, but they couldn't. As a result, Mexican newscasts carried only the most predictable "news" items—what the president said, where he went, who he hugged, what he wore. If a radio commentator criticized the government, that was that—he was out of a job.

But when the peace talks came around, the earth shifted. The radio waves in San Cristobal de las Casas carried the voice of Subcomandante Marcos as he spoke before a press conference after the talks. He chided the government for its treatment of indigenous peoples, for human-rights violations, for lack of social reforms. His voice filled the town square, resounding through homes and businesses as townspeople gathered in absolute silence. In the hush of his pauses, you could hear a pin drop.

Upon my return to Miami, I decided I would not travel for the remainder of my first trimester. After a couple other reporting trips that year, safely into my pregnancy, I returned to Mexico for the presidential elections. By then I was six months pregnant. That trip would bring more than a chance to cover a volatile campaign and the aftermath of the leading candidate's assassination. Unexpectedly, it would bring me face-to-face with my father's past once again.

When I went to interview Ernesto Zedillo, who had become the ruling party's chosen contender after the leading candidate, Luis Donaldo Colosio, was murdered, a young staffer greeted me with a friendly smile.

"Hi, cousin!" went his welcome.

I smiled politely and thought nothing of it. His name was probably Salinas and he was just teasing me. After all, the outgoing president was also a Salinas, and, hey, maybe we were all cousins. But this young man didn't let up.

"No, really. We are related. My grandmother is María de Los Angeles Cordero..." he said.

Oh my God, I thought to myself, *that's my father's sister's name.*

"...and I have a great-aunt, María Elena, whom you were named after. They call her La Bebita," he continued. "And then there's my great-uncle who moved to the United States in the 1940s..."

I looked at him and at that moment he looked just like my father. His eyes were green and prominent, just like my father's. I couldn't speak. I could only stare into them. The spell was broken when Zedillo walked in, leading me away for our interview. As I asked him my prepared questions, my thoughts drifted back to this "cousin." I don't remember what the candidate and I talked about that day. I'm sure I asked him about the death of Colosio. But whatever I asked Zedillo, it had nothing to do with what I was actually thinking.

After the interview, I walked back to where the young man was sitting and we chatted a while. His name was Fernando Solis Cámara. (As it turned out, he was Zedillo's image consultant, which was a formidable task because the candidate was not the most charismatic guy in the world.) I asked him about my family. He offered to set up a dinner with some relatives.

The following night I found myself at a table with about fifteen relatives, most of them strangers. I recognized two of them as the son and daughter of my aunt María Elena, one of the two sisters my father had stayed in touch with after he distanced himself from the larger family. The daughter was a sweet-faced woman named Martita Palafox, and the son was an opera singer named José Luis. There was also another cousin I remembered from my childhood. And then there was the quiet older woman seated at my side. She barely spoke. She only smiled and stared at me with such endearment it made me nervous.

"*Esa es La Muñeca*," one relative announced after a while.

La Muñeca? I couldn't believe it. It was my cousin María Elena, whom we lovingly called "The Doll." She had been my father's favorite niece and we used to visit her in Ensenada when I was a little girl. She used to work with my uncle, who, it was said, had discovered a cure for cancer. La Muñeca was the keeper of his secret formula. They said she kept the composition in her head and never wrote it down. But long after I'd seen her last, she had a stroke that left her paralyzed for quite a few years. I had no idea she had improved to the point that she could walk. Still, she could speak only with great difficulty. She took out a small, tattered telephone book, held together with a rubber band, and opened it to reveal old photographs of my family. There was my father, my mother, my niece Cici, my nephew Charlie. She pointed at the pictures insistently. She wanted to know how they were doing. I hadn't even recognized her, yet there she was next to me, asking about my family and smiling as she gazed at my belly. Another cousin on the way, she must have thought. It was so wonderful to see her. She brought back so many memories of my childhood. At her side, I felt like I had finally found my long-lost family. It was an overwhelming feeling of joy.

My newfound family was thrilled about my pregnancy. They wanted to know all about my life, my husband, my job. They made me promise I would bring the baby for a visit. After dinner, my cousin, Zedillo's image-maker, gave me a list with all their names and telephone numbers, so I could keep in touch.

Before I left Mexico, I asked my cousin if he knew any-thing about my father's reasons for leaving the priesthood. Did my mother have anything to do with it?

"Well, I know he used to be a priest, but I don't know why he left," he responded. He told me there was another relative who insisted my father had made up his mind to leave the priesthood before he met my mother. But who was this rela-tive and what did he or she know? I now had more questions than I did before. But I also had more relatives than I did be-fore. Perhaps one of them held the key to the Box of Secrets that was my father's past. I vowed I would keep in touch.

That November 1st I gave birth to a beautiful baby girl with plump cheeks and sweet, tiny feet I'd recognize any-where. Julia Alexandra Rodríguez. I could not stop crying from the moment I saw her and held her in my arms. She was my dream come true. And the world as I knew it would never be the same. I became not only a mother, but a mother first. My love affair with my job became a secondary fling. Eliott and I were ecstatic. Julia's birth would bring us closer than ever.

With two daughters from a previous marriage and a new baby, Eliott's paternal needs were more than satisfied. But I was just getting started. The following year, I got pregnant again. I was so excited that I put my positive home pregnancy test in a gift box and wrapped it. I invited my husband out for a romantic dinner, and at a seemingly appropriate time I handed him the surprise box.

But when he opened it, his jaw dropped.

"Oh, no."

It was not what I wanted to hear. But I knew that sooner or later he'd be as happy with the idea of a new baby as he was with our Julia.

Not too long afterward, I traveled to Chicago to attend the awards banquet at the National Association of Hispanic Journalists' annual convention. I was going to emcee the awards gala, as I had for several years. I was feeling perfectly fine. But the night before the event, I felt a familiar wave of cramps. I knew exactly what that meant: another miscarriage. I was heartbroken.

The following night, I stood at the podium before hundreds of my fellow journalists, emceeing the awards banquet, trying desperately to hide the pain. I felt alone in my grief, and I was angry at my husband for his reluctance to have another baby in the first place. It was an irrational thought, I know. But I felt I needed someone to blame; then maybe the loss wouldn't hurt so badly this time. Only, Eliott felt that loss, too—he was just as devastated as I was. And perhaps it was this shared sense of longing for the three angels that had vanished from our lives that brought us together, on a warm night in August 1996, in the powerful instant that

Gaby was conceived. I gave birth to my beautiful Gabriela María on May 9, 1997.

MOTHERHOOD BECAME the lens through which I would come to contemplate the world. It became increasingly difficult to leave my daughters at home while I went away on assignment. Everything about them crept into my time away—their voices, their tears, their pictures. Yet having them also gave me a sense of empathy for the stories of others, a sense I don't think I had felt until then.

SEVEN

Everyday People

The camera can be a cruel beholder. It sees everything, even the tiniest blemishes. Like it or not, if you're on TV every night, you are likely to become a makeup expert. You learn to wield concealer like a magic wand against the shadows of fatigue, or last night's lack of sleep. But other things aren't so easy to hide. Your feelings, for one. Disguising them takes a lot more than a good concealer.

I know the rules of journalism, particularly the unwritten one that forbids reporters from getting too involved in the stories they cover. Don't get involved politically, of course. Don't get involved socially. And don't get involved emotionally. But it is that last part that I often wish I could cover with makeup. I confess, I've broken the "emotional" clause more than once. I have not only opened my heart in the course of reporting, I've left it behind on foreign roads and in the homes of strangers whose names I don't always remember. But how does one walk away from a grieving woman who reminds you of your mother? How does one simply leave a distressed family that could be your own? How does one cover up the feelings stirred by stories of loss and love? Oh, yes, I've mastered a nifty trick to dry up my tears in a hurry, during breaks on the anchor desk. I just lean my head back and blink the tears back into my eyes. It works—or at least it prevents the mascara from running. It doesn't exactly turn your heart to Teflon.

I know I've broken this rule because there are some stories that have lingered for years in my memory. These are not big journalistic scoops. It's not the breaking news angles that make them so memorable. It's the human element, the way these "sources" taught me valuable lessons about the resilience of the human spirit. I share them with you just as I remember them:

EL MARIACHI

I traveled to Mexico City at the height of the economic crisis of the mid-1990s to work on a preelection, one-hour special called *Mexico in Search of Its Destiny*. The idea was to see the country not through the eyes of politicians, analysts, or intellectuals, but through the eyes of its people. My topic was the economy. It was a time when everyday people struggled against sky-high interest rates and unbelievable levels of inflation. They were losing homes, cars, jobs. If they had taken out loans, the erratic interest rates had tripled the amount of their payments, sending them into a spiral of despair. So we set out to depict the hardships faced by the average Mexican citizen. We began our search at Plaza Garibaldi, the gathering place for Mexico's most recognizable cultural icon, the mariachi musician.

The festive square, dotted with noisy cantinas, was bustling—in ironic contrast with the larger economic depression. Mariachis, dressed in their crisp, ornate suits, strolled and sang for tips. One of them in particular caught my eye as he crooned a sad *ranchera*. I approached him and asked how he was coping with the crisis. He said he was struggling to make ends meet. He was married and had two daughters who attended a trade school for computer training. He was afraid he might not be able to pay for their education much longer. I asked if we could interview him at home with his family, and he gave me his address.

The next day, my crew and I wove through the hilly outskirts of the capital, searching the boonies for this mariachi's house. We traveled up a steep road until the pavement ended and the road stretched into a dirt path. We found his house at the top of a hill. It was a curious structure with walls of exposed brick and a roof that covered only half of the house. I knocked loudly on the front door, as there was no bell.

"¿*Sí?*" a woman called out to us from the other side of the door. I asked her if the mariachi was there. No, she said, he was out. "My husband had to leave. He couldn't call you because we have no phone," she hollered.

So we asked if we could interview her.

"No, no, no," she replied.

"Please, just for a few minutes," I begged.

"I'm not dressed properly," she argued.

"We'll wait for you," I came back.

A few minutes later, the door opened onto a humble country house. The mariachi's wife led us through patches of light and shadow—the effects of the partial roof—in the small living area. In the rear, I could see wired coops of small farm animals and hens pecking about in the general area. The woman told us she was a nurse but wasn't working full-time because it was very difficult to find a job. She did only odds-and-ends nursing work. In fact, while we chatted a neighbor came by to tell her a child was very ill next door. So the mariachi's wife excused herself and rushed out to administer an antibiotic shot. That's how she contributed to the household budget, she explained when she returned.

As she spoke, she led me into the tiny bedroom she shared with her husband. It was a spartan room with only the barest of furnishings. But the small bed glowed luxuriously. On it she had laid out his mariachi suit for him. It was so beautiful and elegant, like a precious relic, the suit of a king. And that magnificent hat, symbol of Mexican pride. The entire outfit seemed jarring against its faded, worn-out surroundings. Amid all that poverty, here were the golden clothes that turned him into royalty for a few hours each night.

The mariachi's wife then led me out to a jagged-topped wall in the main house. The reason for the wall's unevenness was that they were building it brick by brick. Every time they raised a little money, they would add more bricks.

"Little by little, one day we'll have four walls and a roof," she explained.

Her daughters breezed into the room, their hair still damp from the bath. They kissed her goodbye and took off for school. They would walk down the dirt road to catch the bus, she explained. And at night they would climb back up. I found that to be a startling thought, considering the hill below the house was very steep and long. It never fails to shock me that Mexicans can live this way, considering Mexico is an oil-rich nation that is home to some of Latin America's wealthiest families. But at the mariachi's home nobody was complaining. The girls considered themselves fortunate that they could go to school. And they did so thanks to the torrid love songs their father sang each night. Each ballad meant another day in class, another family meal, another brick for the jagged wall. It was a stark contrast: the festive tunes of Mexico's folklore concealed a common struggle for survival.

We never got to interview the singer at his home. But he was there, in that suit, its glowing promise filling each crevice of the house.

A PERUVIAN ELECTION

Soon after I picked up my first TV news microphone, I began covering elections. I've covered all kinds—municipal races, congressional races, presidential races. I've interviewed people who could vote but didn't, people who wanted to vote but couldn't, and people who didn't even know there was an election going on. But for all the stories I've done on voting, the one that remains in a category apart is the one I heard from a poor Peruvian widow during the 1995 reelection campaign of Peruvian president Alberto Fujimori.

We found her on the outskirts of the capital as we scoured the city for voter reaction and stories to lend a human angle to the political coverage. She was a poor woman who had moved down from the mountains after the Sendero Luminoso (Shining Path) guerrillas murdered her husband. She had several children, and they all lived in a small shanty of dirt floors and sheet-metal walls. They lived in horrible conditions and with such scarcity that the children were filthy and barefoot. They slept on a couple of thin cots and cooked their meals on a tiny open stove that held a single pot.

SHE WAS an Indian woman who spoke only rudimentary Spanish, but it didn't take long for her to convey to us the tragic nature of her story. She represented a poor, indigenous population caught between the rebels and Fujimori's army: the rebels used them as camouflage, which only made the army hunt them down as guerrilla sympathizers. After her husband was killed, this woman brought her children down from the mountain and together they scratched out some kind of living. Unfortunately, not even moving away from Sendero's path spared them of suffering. Her oldest son, the one she relied on to work and support the family, suffered grave injuries after the bus he was riding was attacked by Sendero rebels.

The young man wound up at a local hospital, but the mother had no way to pay for his treatment. So she pawned the only thing of value that she possessed: her voter registration card. She had surrendered it at the hospital as payment for her son's care. If she wanted it back, she'd have to pay the bill. Of course, the downside was she couldn't vote. And for her this was a tragic thing. She wanted desperately to vote in the presidential election. Her life, the lives of her children, all depended on her vote, she insisted.

"Who would you vote for?" I asked. Her response was adamant and unequivocal.

"El Chino." She meant Alberto Fujimori, an agronomist who descended from a modest Japanese immigrant family. The hardline, authoritarian president who took Peru by storm was her candidate. He had set out to dismantle Sendero Luminoso—and he eventually did. Plus, she had seen him in the countryside, tossing pencils and school supplies to poor children, from his "Chino-mobile"—a flatbed truck fitted with a special protective railing. She took this as a sign that he valued education for even the poorest Peruvians. And these facts combined to convince her that if she could only buy back her voter registration card, her children would have a chance at a good future. For her this thought outweighed the fact that her family had little food, few clothes, and no floor, that they awoke to worsening misery each day.

None of this made much sense to the skeptical journalist in me; but it deeply moved the mother in me. Her story hit me so hard that I had to halt the interview halfway through. I signaled to Angel Matos, the photographer, and to Marilyn, who was again my producer, to meet me outside. Away from the shanty, I felt tears fill my eyes. When I looked up at my colleagues, my very close friends, I realized they, too, were misty-eyed. We took a moment to compose ourselves and returned to the woman's house to finish the interview. When we were done, my colleagues and I reached into our pockets and gave her whatever cash we were carrying.

"Go buy clothes, school supplies, food for your kids. Forget about your voter registration card," I told her as I was leaving. But somehow I don't think she did.

THE CRUMBLING SEAWALL

Havana is a mirage. At night, it sparkles like a golden crescent against the venerable Malecón, the famed seawall. By day, the majestic colonial facades rise in the Cuban sun, fooling the naked eye with their splendor. Yes, they are splendid in the way sepia-tinged photographs are splendid, as objects to admire from a distance. But when you get close, you realize they're falling apart, piece by piece, and once you get inside you find slums.

On my first reporting trip to Cuba, in 1995, I set out to do a story on the magnificent architecture of Old Havana.

That I had even been allowed to report in Cuba is a whole other story. The Fidel Castro government had banned Univision from the island for many years. Because the network was based in Miami, we were perceived as being one-sided and overly influenced by Cuban exiles. But at the suggestion of my husband, I contacted officials of the Cuban Interests Section in Washington, D.C., and urged them to give me a visa. It was a hard sell, but I finally convinced them to let me travel to Havana without a camera, simply to meet with government officials and state Univision's case there. We did that, and after three days and many contentious meetings, we convinced them to let us return with our cameras.

On my first credentialed visit, one of my many interviews brought me to the office of Havana's foremost historian, Eusebio Leal. I had sought him out after I'd seen all the decrepit colonial buildings in Old Havana and along the Malecón. Out of respect for his errudition and expertise, I let him go on and on about each and every historical nuance of Havana's architectural gems. At the end of a very long interview, I asked him about the crumbled structures I had seen. He answered my question, then gave me a wry smile.

"I don't know why I get the impression," he added, "that you are not going to use anything I've said except for the part about the buildings falling apart."

And he was right.

I went into one of those seaside buildings. Once grand and spacious, it was now a *solar*—a slum. It had been subdivided so many times that families seemed to melt into one another. The walls blistered and peeled in the humidity. Roaches scampered across the muddy floors. Storms and general decay had poked holes in the roof, and water leaked through, soaking clotheslines, possessions, floors.

In this building I found a woman and her baby living amid the squalor. I watched her carry the infant up and down a flight of open stairs. The steps were wet and muddy and there was no railing to steady her climb and descent. The room where they slept was on the upper level, so she had to climb those stairs several times a day. Their living conditions were unfit for human habitation. There was no electricity, no working toilets. The baby's nose was runny and she was cranky. I couldn't take my eyes off the child. At the time, my oldest daughter, Julia, was one year old, barely older than this baby, who, despite the extreme poverty, had clothes and shoes to wear.

"I have family in Miami," said the mother. "They send me clothes. I sell most of them so I can make a little money to survive."

But how on earth could she survive there? How could she live, day in and day out, in such conditions with a baby? The image of them ascending those stairs is burned into my memory. All I could think about was my own baby, how fortunate she was to have the smallest blessings we often take for granted. Once again, I reached into my pocket. I gave the woman whatever I had, no more than fifty bucks. As a journalist who was getting to know Cuba's jagged edges, I knew that my money and my sympathy would make no difference at all in her life; but as a mother, I had no choice.

CITY OF MUD

In January 2001, as Washington, D.C., prepared for a presidential inauguration, a massive earthquake rattled El Salvador, unleashing mudslides, toppling mountains, wiping out neighborhoods, swallowing families. I've covered my share of natural disasters and have witnessed the devastation they've caused. But in most cases I've parachuted in, scrambled for interviews, and rushed out. This is what you do—report, write, edit your story, and move on to the next assignment. But San Salvador hit me hard. Getting there was nearly impossible. We had to fly to Tegucigalpa, rent a small plane, and fly into a Salvadoran military airport. Getting around the devastation was even more difficult. Roads had been wiped out. Mudslides had flattened mountain neighborhoods. In some places it seemed as if the entire city was under a thick layer of mud. Only broken power lines and crooked antennae poked through. The quake had hit on a Saturday morning, so children were home from school. Their school was spared in the quake. Their homes weren't. Tragically, many of them were missing in the muck and rubble.

Besides covering the quake aftermath for the newscast, my crew and I were putting together a story—the anatomy of a rescue—for Univision's newsmagazine, *Aquí y Ahora*. Once again, Angel Matos was my photographer and executive producer Patsy Loris was commanding our team. We made our way to one of the mountain neighborhoods where residents were attempting to dig out their loved ones and what remained of their homes. We walked carefully along the damp ground of what had been a working-class neighborhood. Buried below our feet were homes, trees, cars, people. Amid the devastation, we found a man and woman digging through the rubble of their home, searching for their two children and their baby-sitter. For hours, they dug and dug through mud and crushed brick. They dug against the clock, knowing that before long the bulldozers would come to clear the zone. They dug desperately but warily, careful not to stab the surface too harshly. As dusk fell on the excavation site, they continued to dig, illuminated by our photographic lights, until they could stand no longer.

The following morning I found the father back at it. He

was wearing the same clothes. Clearly, he hadn't slept or eaten. His wife had collapsed and had to be taken to one of the shelters, so he dug alone.

"I don't think I'll find them alive," he said, "but I have to try."

He and his wife had been working when the earthquake hit. They rushed home to find everything gone, buried. The pain of his loss etched deep lines in his face as he squinted against the sun. We watched as he dug for hours into the night, until he had to rest.

On the second morning after the quake, we found him in the same despair.

"I don't think I'll find them alive," he said again. "But I want to find their bodies intact. I don't want the bulldozers to come and mangle them."

We left him to cover other stories, and when we returned we found him still digging. He had found some school supplies, his daughter's little purse, some toys. He was filthy, smelly, but determined to find his children. He worked in an area where the bulldozers had not yet reached. Families there all seemed to be in the same predicament, scrambling to find the missing. The stench of death rose from the ground. Some people pressed surgical masks over their mouths. Others threw up. Every once in a while, you'd hear a shout:

"There's a body part over here!"

Or, you'd hear: "Woman. Long hair. White blouse. Lace on the collar."

And someone would reply: "That's my daughter."

Rescue workers would take the body and place it in a bag with an identity tag. The families who were digging there used little shovels, like the kind that come with children's beach buckets. They dug delicately, meticulously, like archaeologists searching for ancient remains. But time was running out—the bulldozers were drawing nearer by the hour. Our own deadline was drawing near. We were scheduled to leave El Salvador the following morning, and we still had to write, edit, and transmit our story. As night grew darker, I nudged our photographer, Angel, and motioned that we had to leave soon. He was exhausted, emotionally drained. But he shook his head and kept shooting as the man kept digging.

"I can't leave," he whispered. "Mine are the only lights he has. How can we leave?"

He had gone through several batteries. Each time one ran out, he'd go to his bag and get another one. He was no longer shooting for our eight-minute story. He was keeping the light on for the buried children.

Then, all of a sudden, the father we were following started yelling:

"¡Aquí está! ¡Aquí está!"

Rescuers rushed over to the mound of dirt where he was

standing and uncovered a child's leg and foot. The father crumbled to the ground.

"That's my daughter," he wept. "I can stop searching. This is where it ends for me."

I asked him how he knew it was his daughter. He said something that has stayed with me.

"A father knows his daughter's feet. A father knows his daughter's toes."

I thought of my own little girls and the way I tickle their toes and tell them they've got the prettiest little feet in the world. I had no words. Angel and I picked up our stuff and left the mountain. We rode in silence to the office, where we spent the night putting together our story, still numb from what we had seen. We left El Salvador the next morning, but I couldn't shake the images. I tried to return to my daily routine, even keeping a manicure appointment. But as soon as Maribel, the manicurist, asked me about my trip, I burst into tears. And when a producer complained my story ran a minute too long, I replied to her:

"Okay, cut it. But you make the decision on what you want to leave out. The part where he's digging desperately? Or maybe the part where he finds his daughter's foot? You decide."

For the life of me, I couldn't understand how anyone could fuss over one minute of such a tragic story. Then again, I wasn't back in TV mode yet. I had spent days following this father's plight and was still there. I poured all my memories into my syndicated column, which I titled "Walking upon Death."

Just a couple of days later, Angel and I flew to Washington to cover the first presidential inauguration of George W. Bush. Taking my seat on our rooftop set overlooking the White House, I felt horribly detached from the story. The city below swirled in pomp and festivity. In some ballroom, the president's daughter Jenna wore a strapless gown as she danced with her father. Revelers in tuxes and western boots filled the inaugural parties. A variety of images streamed across the monitors on our rooftop set. They were lovely and elegant. My God, I thought, what world is this? Here I am in the most powerful, most glamorous, few blocks on earth and my thoughts remained fixed on a patch of mud thousands of miles away.

About a month later, in Miami, I got a letter from a friend of the Salvadoran family we had followed. They remained, I was told, homeless and childless, left to rebuild their lives. But through this friend the family conveyed their thanks for our airing the story. In a poignant postscript, the friend added that they had found the body of their other child, a son, shortly after we left. I wondered if they had had to wait for the morning light.

EIGHT

❧

Dictators, Strongmen, and Comandantes

There are the ordinary people and there are the freaks of nature, the creatures that thrive on absolute control over their environment and the humans who inhabit it. Although they may differ in ideology and methodology, they are joined in a rogues' gallery of Latin American strongmen. And they share some traits. They live in a bubble. They surround themselves with layer upon layer of bodyguards, yes-men, screeners, attack dogs, lap dogs, and various other lackeys whose sole purpose in life is to convolute matters for the outside world so that their masters remain undisturbed in their fancies. Their entourages rival even that of J.Lo herself.

In one memorable year, 1989, I interviewed Chile's Augusto Pinochet, Nicaragua's Daniel Ortega, and Panama's Manuel Antonio Noriega—a right-wing dictator, a left-wing wanna-be dictator, and a strongman willing to fly whichever way the wind was blowing.

I can't say I know what life is like inside their bubble, but I've seen glimpses. And from what I've seen, not even García-Márquez could have conjured up such a cast. At least not without the help of Fellini.

NORIEGA'S CRUSH

General Manuel Antonio Noriega, Panama's feared military strongman until a U.S. invasion settled his fate, landing him in a Miami federal prison, wasn't fond of television interviews.

Truth be told, for all his power and his "Doberman" guardsmen, he seemed to suffer from a bit of social anxiety disorder. When Noriega failed to respond to countless requests for an interview in 1989, just months before the invasion, I asked our Central America correspondent why there was no word. The correspondent, Monica Seoane, let me in on one of his quirks: "He doesn't like to be approached by people he doesn't know."

Monica wasn't trying to be sneaky or overly protective of a special source. Sure, she had Noriega on speed-dial. But then again she had confidential sources all over the region, on all sides of the political spectrum. And if she had wanted to protect an exclusive source on journalist relationship in this case, I would have understood. But that was not the case. In fact, she seemed personally exasperated at Noriega's refusal to respond to our interview requests, one way or the other. She promised to help however she could.

And so when Monica heard there would be some kind of event outside the presidential palace, she rang the general and told him she'd be there. "Look for me," she told him.

When we arrived at the event, she told me to wait for her at a nearby corner, behind the press barricade.

"Okay, when he comes to talk to me, I'll just call you over," Monica conspired.

I waited behind the barricade for a sign of Noriega's arrival. And when he finally got there, it was just as Monica had predicted. The general walked right over to talk to her. Who could blame him? She was a sultry, green-eyed free spirit with an easy smile. To his surprise, she called me over and introduced me. I could tell he was displeased. Clearly, he wanted to talk to her alone. Even so, he invited us into the palace for a personal tour. In one telling moment, he called over his president, Manuel Solis Palma:

"Come here, I want you to meet my friends."

After our stroll through the palace, Noriega agreed to give me an interview in the following days. I was in—or so I thought.

On the day of our interview, we brought Monica along to help smooth our path. When we arrived at the barracks where the general had his office, his "Dobermans" split us all up. My producer, Rafael Tejero, my photographer, and our equipment were taken to separate rooms, while Monica and I were led into another room where we were screened.

When the guards finished with us, Monica telephoned the general to tell him we were ready for him. I could see she wasn't happy—she rolled her eyes and huffed.

"No, Tony. I said no. I told you I was not . . . Tony . . . Listen to me . . . I'm not doing the interview. María Elena is doing it, and that's final!"

It was as if she were talking to a bratty little brother, or, worse, to a submissive husband. And, boom, she hung up on him. The general had met his match. She turned to me and cracked a smile.

"Don't worry. You'll do the interview and everything will be fine."

With that, we made our way to Noriega's office. He led us into a small sitting area, and after a bit of casual chitchat, Monica excused herself and left the room. When I saw the look on Noriega's face, I wanted to run and drag her back in. The room seemed so empty without her. It was just the general and me and a shock of silence. Noriega moped as I glanced around, noticing the oddest details, like the wooden rocking chair with the name "Tony" carved on its back. In the background a TV flickered. Noriega stared at it, twiddling his thumbs, until he finally broke the ice:

"Look—there's your network. I watch it all the time."

I turned around to look at the TV. It was tuned to Televisa.

"I don't work for Televisa. I work for Univision," I explained, straining to make any kind of small talk. I tried to drag things out as best I could until it was time for our interview, which we agreed we would do outdoors, next to the Panama Canal. We were going for a more natural, casual setting. Unfortunately, no amount of fresh air put the general at ease. His answers were uniformly stiff and cynical.

"If the people want you to leave, why don't you?" I asked him.

"Because the problem is not a man, it's a canal," he replied.

"When will you step down?" I insisted.

"When the time comes," he shot back.

"When will that be?"

"When it's time."

I know he was probably hoping that it was time for me to go. His time to go came just a few months later.

FUJIMORI, BY COVER GIRL

I interviewed the former Peruvian hardline president Alberto Fujimori twice, once in 1992, during the Summit of Andean Nations in San Antonio, Texas, and the next time some years later, when he ran for reelection for a third term. Luckily, during that last interview he didn't recall our first meeting.

Our first meeting got off to a rocky start and ended even more disastrously. He had walked into the interview room with a cocky attitude and a bizarre question:

"Do you have any makeup?"

Makeup? I handed him my compact. He opened it, then peered into the little mirror as he dabbed the powder puff all over his face. Shineless at last, he took my questions. And I had a lot of questions concerning accusations of corruption in the Peruvian armed forces and police.

"There's no corruption in the military," he snapped.

I listed a few examples for him of the accusations we had heard. He argued that such accusations might be true for the police forces, but not for the military.

"So you admit it?" I asked him.

Not exactly, he said, launching into a justification.

"Then you're justifying it," I said.

"I'm not justifying it. I'm saying there are reasons," he replied.

We moved on to other topics, but he was still irked when we finished our interview.

"I don't like the way you asked me that question," he told me. "You made me sound like I was admitting and justifying corruption."

I tried to explain politely that I had given him ample opportunity to state his case.

"Yes, but those accusations were unfair," he insisted.

I told him I had based my questions on a document from a Peruvian source.

"Well, then, ask me again, but this time ask me about that specific person," he pressed.

"No, sir," I told him, "I'm not going to set up a question like that."

I was not about to allow him, or any other source, to tell me what questions to ask. But my producer, Marilyn, urged me to give in.

"Just do it, María Elena. Just ask him the question," she told me.

And so I did, knowing full well that I was not going to use it in the end.

A few years later, I traveled to Peru to cover his reelection campaign. We arrived on the red-eye and went directly to the military airport to accompany Fujimori into the countryside. During the flight, I decided to venture from the press area and wander to the front of the plane. I wanted to remind the president he had agreed to a one-on-one interview. But when I got up there a guard stopped me before I could reach the presidential row. I realized Fujimori was stretched out, taking a nap.

"Don't bother him while he is resting, and don't even think of filming him while he's sleeping," the guard warned. So I backed off, realizing that I had caught him during his most precious time of the day, nap time. Fujimori wasn't one to sleep at night. No, he stayed up late, reading, surfing the Internet, shooting off emails. Night wasn't for sleeping, it was for working. Only those random, stolen hours, such as those spent aboard an airplane, were for sleeping.

On the ground, we rode with the president atop his "Chino-mobile," his flatbed truck, roaming poor, indigenous areas. His staffers would toss campaign souvenirs to the gathered masses and people would swarm over to grab whatever they could, a pencil, a T-shirt, a baseball cap. From his perch behind the flatbed's railing, Fujimori seemed to delight in the mob scene below. He chuckled at the sight of people trampling one another, children falling, hands reaching.

I clutched the railing, horrified at the sight, and even more so at the president's disregard for the safety of his people.

"Aren't you afraid someone could get killed?" I finally asked him.

"No, it's nothing. These people know what to do," he replied, waving to his fans. "They love this."

Still fresh from his restorative nap, Fujimori beamed a candidate's smile. I, on the other hand, was a mess, having gone nearly twenty-four hours without a wink of sleep and now braving this white-knuckle ride. And I wouldn't get my one-on-one interview until 1 A.M. I wondered how I would pull myself together for my stand-up. Maybe the president could lend me his compact.

THREE PRESIDENTS IN THREE DAYS

In February of 1997, the president of Ecuador was overthrown by the Ecuadoran legislature on grounds of "mental incompetence."

The coup made for a big, colorful story, the kind any journalist would rush to cover. But I had a hard time rushing anywhere in those days. I was seven months pregnant with my second child. The last thing I needed was not one but three "presidents" to chase.

You see, in the chaos that followed the coup, three of Ecuador's top politicians claimed the constitutional right to be president. The first, of course, was the newly deposed president himself, a flamboyant character and sometime singer and comedian named Abdala Bucaram, a guy so obstinate that it took more than one coup to keep him out of Ecuador. Then again, he called himself "El Loco."

Then there was his vice-president, Rosalía Arteaga—she believed the law favored her right to be interim president. And, lastly, was the head of the national congress, Fabian Alarcón, who insisted that according to the constitution he should claim the honor.

And so, when my crew and I landed in Quito, we weren't sure who to interview first. Was it the president behind door number one, two, or three? We covered our bases by interviewing both the head of the national congress, who had led the "mental incompetence" charge, and the vice-president. It was the latter who gave me a cell phone number for Bucaram, who was stewing at home in Guayaquil. When I called him and requested an interview, he agreed.

We flew to Guayaquil and went straight to Bucaram's house, where he gave a press conference. But he riled himself up into such a state that after he finished he stormed into another room and shut a glass door behind him. I followed him and tapped on the glass.

"You said you were going to give me a one-on-one interview," I told him.

"No. I already said what I have to say," he snapped back.

Then I saw his eyes light up, and I turned to find a tall, voluptuous blonde in a miniskirt, a magazine writer from Chile, standing behind me at the door. She tapped on the glass and purred a compliment.

"You're looking so slim, Abdala," she said, then asked him for an interview. To my indignation, he agreed.

"Come back in a little while," he purred back, and the next thing I knew he was out of sight.

There I was, seven months pregnant, feeling unglamorous at best. But I was not going to let some sex kitten steal my interview. My deadline was approaching. Soon I would have to go back to the local affiliate to write, edit, and feed my story for the newscast. That's when I did one of the most demeaning things I've ever done in my career. I tapped on the glass again, and when he peeked out I pressed my belly to the glass. I didn't recognize the alien voice that came out of me.

"I didn't come all the way here and risk the life of my baby to leave with no interview!" I thundered.

I simply was not going to take no for an answer, certainly not from El Loco. A good part of my career up until then had been built on interviewing the presidents and leaders of Latin America. This guy was not going to be an exception.

Bucaram's wild eyes narrowed on my belly.

"Come back in two hours," he said.

So I went to the local station and put together my story and rushed back to his house. But when I got there, they wouldn't let me in. I pleaded with a guard.

"But he personally told me to come back," I insisted.

"Sorry," said the guard.

So I grabbed my cell phone and called Bucaram. I called him again and again, until he picked up. Once again, the alien emerged:

"I'm standing outside your door. You said you would do this interview and now your guards are telling me I can't come in."

There was a long pause. Then:

"Okay. Come on in," said the president.

Moments later, he was railing against his political enemies, vowing to make his return. I couldn't help but remember the last time I saw him. It was during a previous coup, when he was forced into exile in Panama. I had spotted him by the swimming pool of our hotel one day, lounging in a Speedo. The memory jarred me back into our interview.

"I'll be president five times, and five times they'll overthrow me, and five times I'll return!" he fumed.

Frankly, it wouldn't surprise me, but it didn't matter. The

important thing was that he was speaking into the microphone. At last.

THE PALACE OF PINOCHET

When Chilean dictator Augusto Pinochet held a plebiscite on his rule in 1988, a vast majority of his compatriots sent him a strong message: step aside and bring on the elections.

Their cheers filled the streets outside the Moneda Palace in Santiago. From the rooftop terrace of the Hotel Carrera, I could see the celebrations and hear their chants:

"¡Y ya cayó! ¡Y ya cayó!" (And he fell! And he fell!)

A year later, I returned to Santiago for the elections, with one goal in mind: to interview Pinochet. It was an ambitious goal—the general didn't give many interviews, particularly to the foreign press. But our producer, a Chilean-born journalist named Francisco Ginesta, had good contacts in the cabinet. One of the ministers was an old friend from college. So Francisco called him up and asked him to help us get the interview.

I wasn't too hopeful, but one night Francisco showed up at my hotel room with a bottle of nice Chilean champagne.

"Tomorrow we interview the general," he announced with a grin.

I didn't believe it, but I drank the champagne anyway. Then I prepared meticulously for the interview, just in case he was right. I read news clips and composed a long list of questions, which I edited and reedited. I picked out my interview outfit, a bright crimson dress with a matching bolero jacket, gold buttons, and shoulder pads—my "general suit." It was the only time in my career that I prepared fully for an interview not believing I'd actually get it.

It wasn't until we were walking through the ornate doors of La Moneda that I suspected the interview was really going to happen. Palace guards separated us for screening. They took me to a huge room with sumptuous decor and gilded walls that seemed more like beautifully framed art panels. There, an official interrogated me in a way I had never been interrogated before:

What are you going to ask him?

Why are you going to ask him that?

Do you think it's wise to ask him that?

Of course I didn't give them my real questions, just enough to satisfy their curiosity. When they finished, they brought in Francisco and our photographer. Then, magically, one of the gilded walls opened into another room. I expected to see the Wizard, but instead there was Pinochet, seated on a kind of film set, illuminated by bright lights and flanked by a dozen men in military uniforms. There was one empty chair.

Mine, I supposed. I had never seen anything like it. The general not only brought along a few trusted guards, he brought along his own video crew. We had to haggle with his press people to be allowed to use our own cameras and microphones. We were not about to let them take control of our interview. By the time we set up, there was a tangle of equipment and microphones pointed at Pinochet. So much for our intimate one-on-one.

Pinochet gave me a curious look. He had the eyes of a sweet old grandfather. Who could have imagined that behind that inoffensive smile lay a ruthless dictator linked to so much bloodshed?

"¿Y tu eres cubana?" he asked. (Was I Cuban?)

"No, I'm of Mexican descent," I replied.

Baffled, the general turned to one of his people. "I thought you said they were Cubans from Miami," he said.

No wonder he had agreed to give us the interview. He assumed that we were (a) Cuban exiles, and (b) sympathetic to his extreme right-wing cause.

I scored even fewer points when I started with my questions.

"Mr. President, you have been accused of violating the human rights of your people..."

"What?"

"Human-rights violations. Disappearances. Murders..."

"What? In Mexico they have more human-rights violations than we have in Chile."

After that, whatever I asked Pinochet, he would answer mostly in uninspired monosyllables:

Sí. No. Sí. No.

It was one of the worst interviews I've ever had. The best thing about the interview was that we got the interview.

"Do you have any intention of running for president in the future?" I asked. He stared me down.

"Do I look like a fortune-teller to you?"

No, he didn't. The glaring light bouncing off the buttons of his uniform revealed the steel blue of his eyes.

FIDEL TEETERS

After so many years of living in Miami, I have, perhaps by osmosis, come to anticipate The Day. The streets will be filled with celebration, the champagne will flow and the airwaves will buzz with the euphoria of a vindicated exile population. The Day—The Day Fidel Falls—is an expectation engraved in the minds of more than one generation of Castro watchers. How will he go, with a bang or a whimper?

But sometime before that happens I'd like the opportunity to interview the Cuban dictator. I know exactly what I'd ask him. I've had an extensive list of questions ready for more than a decade, and I update it regularly, just in case I get the call that the aging "Comandante en Jefe" is ready for his Univision close-up.

I've been writing letters to Fidel Castro for more than ten years, requesting a sit-down interview. I've tried any number of approaches—both official and nonofficial. I've asked his ministers, his consuls, even his relatives. But, so far, nothing.

In February of 2002 I came close. I came face-to-face with Castro himself during a trip to Havana. I had traveled to the Cuban capital to cover the visit of Mexican president Vicente Fox. My credentials gave me access to all the public events related to the state visit, as long as I stayed in the roped-off press area. But Cuban security officials had made the ground rules clear: no questions for the *comandante*. No interviews. No casual banter. We were permitted to speak to Castro only if he spoke to us first. Still, I grabbed my list of questions, just in case.

It didn't hurt to have friends in the Mexican delegation. On the first day of Fox's trip, they managed to slip me into a private reception honoring Havana historian Eusebio Leal. He was receiving a special recognition from the Mexican government, and Castro was expected to attend. Cuban state security agents caught on to my presence and tried to hustle me out of the reception, but I stalled and somehow eluded them long enough to disappear in the crowd. Then Fidel made his entrance.

I wove through the crowd until I found myself four feet from the Bearded One. The ceremony was drawing to a close and guests had begun to shuffle out. I followed Castro as he headed for a stairway exit.

"*Buenas tardes, Comandante,*" I said, catching up to him. He turned, slightly off balance. I quickly joined him down the long, winding stairway, watching as he grasped a railing to aid his descent. He walked not with the gallant clip of a legendary revolutionary icon but with the crooked gait of a random, elderly man.

"Mr. Castro, I have been trying to interview you for ten years," I said.

"I just gave an interview to Mexican television," he replied.

"But I am not with Mexican television," I insisted.

Castro reached for the press credentials dangling around my neck. He wrinkled his brow.

"Where are you from?" he asked.

"Florida," I replied.

"Why are all the beautiful women from Miami?" he remarked in a way that caught me off guard. I meant to reach for my list of questions but found myself a few beats behind my intention. I guess that was the point. Besides, launching into my questions wouldn't have done me any good at the moment. My cameraman was nowhere near us. He was stuck in the roped-off area. Not even his zoom lens could have captured my exchange.

"I'm with Univision, which is watched by millions of Hispanics throughout the United States and several Latin American countries," I said. "And I would like to do an interview with you."

From the corner of my eye I could see his guards approaching. I had to think fast.

"I know your sister, Juanita," I volunteered, reaching for a point of connection. That certainly caught his interest. He stopped and pulled me over to the side of the stairs.

"Do you have a message for her?" I asked him. I told him I buy my medicine at her pharmacy in Miami.

He grew nostalgic for a moment. He had not spoken to his exiled sister in forty-two years. She has been one of his most visible critics, but that's not what he seemed to ponder as he steadied himself on the stair railing.

"She made her choice in life, but I'm not resentful of it," he said.

When we reached the bottom of the stairs, President Fox approached and Castro and he began to walk off.

"What about my interview?" I called out after him.

"What for? They won't allow you to air it anyway," Castro said before his guards finally stepped between us.

My questions were burning a hole in my purse. I was no closer to scratching them off. But I had gotten closer than most. When I returned to Miami, I couldn't say I had seen Fidel fall, or even stumble. But, for what it was worth, I did watch him teeter on the stairs.

WHO IS THAT MASKED MAN?

The masked outlaws were coming to the capital. Mexico City braced for the 2001 invasion of the Zapatistas. It wasn't war but a historic showing by the elusive rebels of the state of Chiapas, who hoped Mexico's new president, Vicente Fox, would be the first head of state to finally discuss reforms for Mexico's indigenous population.

And, of course, the star of this show would be the most mysterious of the bunch, the rebel leader with the *nom de guerre* of an antihero. Marcos wasn't a self-proclaimed, capital-C *Comandante*, but a lowercase-S *subcomandante*.

Subcomandante Marcos was a master of deflecting attention from his true identity. While he courted the intrigue, he deftly shifted the attention to the larger cause of the impoverished and disenfranchised people he claimed to represent. And, despite the fact that his image had become a brand as ever-present as that of any rock star, he remained largely inaccessible to the outside world. It was well known among journalists that one would have to pass through many layers of *comandantes* to get close to the *subcomandante*. But we tried.

We hired Lupillo, a young man from Chiapas, to act as our go-between with the Zapatista leadership, hoping he would help us score an interview with Marcos. We paid for his food and transportation to follow the rebel caravan to the capital, and we asked him to deliver written interview requests to "El Sub." The rebels relied on local volunteers like Lupillo to help them promote their cause, and so he was able to deliver about ten letters, though we had yet to receive a firm reply.

Once Marcos arrived in Mexico City, we decided to hand-deliver our requests to the makeshift rebel headquarters at a local university campus, the Escuela Nacional de Antropología e Historia. While we got the sense from representatives there that Marcos might be available to us, we didn't know for certain. And as the rebels' stay in the capital reached its final days, we grew increasingly pessimistic about our chances. Still, I convinced my bosses to let me stay a couple more days.

On the night before I was supposed to return to Miami, I called the rebels again. And again. I called them up through 4 A.M.

"Call back in forty minutes," they'd say.

But daylight came and it was time to go. I called them once more and told them I was leaving, but that my request still stood. I asked them to give me twenty-four hours' notice in the event the interview came through so we could make the proper logistical arrangements. I had already packed up and sent off my bags with the hotel bellman, and I was getting ready to leave my room for the last time. I had to catch a flight in a couple of hours so I could arrive in time to anchor the newscast from Miami that evening. As I was about to hang up, the rebel spokeswoman on the phone told me to wait a second. Then I heard a man's deep voice.

"*Hola, María Elena,*" the voice said.

"*¿Quién habla?*" I asked.

"*Marcos.*"

Of course. He had me hanging to the last minute and now I didn't know if I should thank him or curse him.

"I'm sorry for making you wait so long," he said. "We can do the interview tonight."

And so I retrieved my bags, unpacked, and anchored the show from Mexico. After the newscast, I went to the university campus for the interview. This time *I* felt like the rock star. I had the biggest entourage ever. Everyone and their mother, girlfriend, and daughter in our Mexico City bureau wanted to meet the *subcomandante.*

We did the interview on the stage of the school's theater—an odd scene that included the TV people, the groupies, and the masked rebels, who, by the way, hadn't bathed during their entire stay for fear that hidden cameras might expose their faces.

"I wouldn't get too close," Marcos joked.

Keeping my distance, I handed him the microphone and he clipped it on with the familiarity of a seasoned pundit. During the next couple of hours we talked about everything—his struggle for social justice, his philosophy, his movement's political goals, his childhood, his romantic relationship, and life in the jungle.

We began by asking him how he originally wound up in Chiapas, and he told us this story:

"I was drunk out of my mind and I decided I wanted to go see the ocean in Acapulco. So I bought a bus ticket to go see the ocean in Acapulco, but I got on the wrong bus. And I ended up in San Cristobal de las Casas, in Chiapas.

"So I get out of the bus and I look around and I say, 'But I want to see the ocean!'"

Quiero ver el mar.

"I'm still drinking and I get on another bus, but that was the wrong bus, too. It takes me to Ocosingo. So there, I asked the indigenous people where the ocean was," he went on.

Quiero ver el mar.

"And they tell me to get in a car and follow the road into the jungle, that there I'd find Laguna Miramar. When I'm deep into the jungle, I start to sober up and I realize I'm nowhere near Acapulco. I look for the exit door, but I couldn't find it. And it's been eighteen years," he said.

Who is this guy? I wondered. What is he talking about?

"Nice story," I told him. "Now tell me the real one."

He laughed.

"It *should* be true," he said. "Don't you think our movement deserves a story like that?"

He ended the fable right there and started the interview. He had allowed us a glimpse into his world of fantasy. But he never let us peek behind the mask.

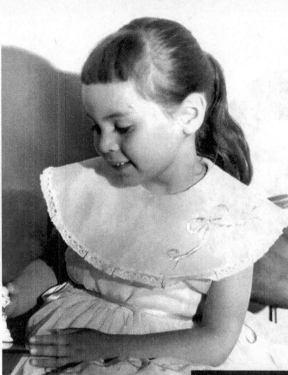

ABOVE: This picture of me was taken by my father. The chair, table, lamp, and porcelain cat were our only furniture items. The rest had been repossessed.

RIGHT: Me in front of our apartment in Los Angeles at age fourteen. (Casa de las cucarachas ...)

Undated photo of my mother.

My father as a priest in December 1933.

OPPOSITE: My father as a priest with his brother, Father José Antonio Cordero Salinas, and their niece Lucila during her fifteenth birthday mass. May 1, 1943.

Undated photo of my parents before their children were born.

Family portrait, early 1980s. From left to right: my sister Tina, my sister Isabel, and myself. Sitting down are my mom and dad.

RIGHT: My niece Cici and nephew Charlie at three years of age.

BELOW: With my husband, Eliott Rodríguez, covering the pope's visit to Cuba in 1998. Eliott was working at the time for Channel 10, the ABC affiliate in Miami.

My daugthers and I in August 1999.
Julia is five years old; Gaby, in white,
is two years old.

ABOVE: 34 KMEX-TV anchor team, early 1980s. From left to right: myself; my co-anchor, Eduardo Quesada; sportscaster, Jorge Berry (standing); and news director, Pete Moraga.

BELOW: Jorge Ramos and I on the set of Univision News.

ABOVE: Interview with Subcomandante Marcos of the Zapatista Army of National Liberation in Mexico City in 2000.

BELOW: Bill Clinton with our crew. From left to right: cameraman Carlos Calvo; myself; Bill Clinton; Alina Falcón, news director at the time, now senior vice president of Univision; Univision producer Vicky Rivas Vasquez; and cameraman Joe Aguila.

With Hispanic soldiers in Camp Victory in Kuwait, a few days before traveling to Iraq.

ABOVE: Interviewing a Hispanic soldier in Iraq.

BELOW: With cameraman Angel Matos next to what was left of the Saddam Hussein statue.

NINE

❧

"Do You Have Work, Mom?"

Dear Julia and Gaby:

Every night, while you are sleeping, I come into your room, kiss you good-night, and tell you that I love you. I know you are in such a deep sleep that you can't hear my voice or feel my lips on your little cheeks. But it's something I need to do.

I cannot go to sleep without expressing my love to you, without looking at you one last time. Maybe the next morning we'll be at each other's throats.

"Eat your breakfast girls!" I'll be yelling, "You need nutrition to learn in school."

"You are so unfair, Mom," you'll tell me for not buying you a $300 portable DVD player.

But when night falls and you are lying there with your eyes closed, cuddled up in your bed, so serene and cute and innocent, I count my blessings.

I am so lucky to have you as my daughters. I dreamed of you almost all my life. When I was fourteen years old your cousins Cici and Charlie were born. I was just a kid myself, but my sisters trusted my friends and me to take care of their babies. Can you imagine that, babies as baby-sitters? It was like playing dolls. But it worked. We took good care of them. There was an instant bonding that has kept me close to Cici and Charlie all my life, loving them and nurturing them as if they were my own. But they weren't.

Since then I wanted nothing more than to be a mom. It took a lot longer than I expected. And the road to motherhood was not easy. It's funny because so many people think that when a woman has children later in life it's because she has chosen to do so by giving priority to her career, and that only then, once you've fulfilled your career goals, are you ready to have a family. Well, that was not the case for me.

Having a career was something I did while I waited for you to arrive. It was not my choice but the circumstances in my life that didn't allow me to be a mom until almost forty years of age. Now I want to make up for lost time. I want to be your mom, your friend, your confidante. I want to know everything about you, and I want you to

know everything about me. I want to share with you a life
without secrets.

I GREW up thinking motherhood was a piece of cake. Or
as my younger daughter would say, "a piece of pie." And I had
good reason to believe it. It seemed effortless enough to my
mother, my role model. She worked tirelessly, yet never ne-
glected or shortchanged us. She loved the time she spent with
us. My most vivid image of her is that of a smiling woman
who juggled her duties as a seamstress with her obligations at
home. I remember her with a pin cushion around her wrist
and scissors in hand, cutting through fabric on her sewing
table, or seated at her sewing machine, peddling away, a piece
of thread dangling from the edge of her mouth.

She always brought extra work home, whether it was
from the fancy bridal shop in Mexico City or the sweatshop in
Los Angeles. And she always had time to finish her house-
work and fix dinner every night. To this day, I have never
tasted more delicious soups than the ones she cooked for us
from scratch. My mother was not only accomplished at her
duties, she was loving, patient, kind, witty, strong, and com-
forting. I dreamed of the day when I, too, would be a mother.
And I wanted to be just like her.

Wishful thinking.

+ + +

THINGS DIDN'T quite turn out the way I expected them to. Yes, I did inherit her work ethic. I juggle a couple of jobs at a time. I bring extra work home. I'm pretty handy with a needle and thread, although I can't cook for the life of me. But at some point early on in my balancing act, I realized that being a working mom is one of the most difficult tasks a woman could ever aspire to.

Of course it's doable. Millions of women do it every day, either by necessity or by choice. But to successfully master the art of motherhood, you have to morph into Superwoman, physically and emotionally. After experiencing five pregnancies in five years, I thought I had endured the hardest part of motherhood. But I was wrong. It wasn't the labor pains that were the most difficult, it was the growing pains of parenting. Returning to work after maternity leave and having to entrust your baby, your most precious loved one, to a virtual stranger is a nerve-racking thing, to say the least. I wasn't as lucky as my sisters were—they had my mother to help them with their babies. By the time I had my daughters, my mom was in her eighties and was incapacitated by a stroke. She could love them, kiss them, maybe even carry them for a few minutes, but that was about it. She needed a sitter herself.

My mom was a survivor. She underwent triple bypass surgery and suffered a couple of minor heart attacks. She had a pacemaker and used nitroglycerin patches. But she stuck it out valiantly.

When my job took me to Miami in 1991, she came to visit a couple of times that year. The last time, she came for Thanksgiving and she decided to stay a couple of months. I was thrilled. There was always a room in my house for my mother. That Christmas, I went on a previously planned trip to South America with my friend Marilyn. My mother went to Orlando with her best friend and her sister—she loved Epcot. But on New Year's Day, she had returned to Miami after experiencing tachycardia. She went to the emergency room at Mercy Hospital, where doctors kept her overnight for observation. The following day, when Marilyn and I returned from sightseeing in Buenos Aires, I found several urgent messages from Miami waiting for me at the hotel. The news was devastating. My mother had suffered a major stroke overnight at the hospital. I don't think I remember ever feeling such a sense of helplessness. I was so far away from her. All the flights were booked for days. We went to the airport anyway. Our correspondent, Osvaldo Petrozzino, brought along a cardiologist friend in case the airline wanted someone to attest for my mother's condition. It took hours, but eventually a couple was bumped from a Miami-bound flight to accommodate us.

When I got to the hospital, I found my mother lying on a bed in the intensive-care unit, motionless. She was completely paralyzed. For days I stood by her side until, little by little, she began to respond, first by opening her eyes, then squeezing my hand. For weeks she lay in a hospital bed, gradually regaining her faculties and motor skills. Her first word was *"cafecito."* She was like a child learning how to walk and talk and respond to sound. She developed a unique sense of humor. She changed around her doctors' names. Her therapist, Dr. Monasterio, became "Dr. Cementerio." Her cardiologist, Dr. Centurion, was renamed "Dr. Cinturón."

But it wasn't easy. She missed her apartment, her furniture, her sewing machine. I would buy her little toy sewing machines, hoping to lessen her anxiety, but it wasn't easy to fool her. Yet she would never be the same. She was no longer the independent woman who took care of herself and traveled as she wished, let alone take care of babies.

So for my girls I was lucky to find Rosario, a Nicaraguan mother of five who would spend ten years with us as our nanny. Not only did she embrace my babies as her own, she proved to be a valuable support system for my husband and me. She took care of baby Julia one fall when Eliott and I vacationed alone in France. It was a much-needed getaway for us. But even as we took a barge cruise along the canals of Burgundy, dining on exquisite food and enjoying spectacular wine, I was miserable. I missed my baby every minute of the day. Each morning, as soon as I could leap off the barge, I'd rush to the nearest pay phone and call home. I would talk to Julia as if she could actually understand what I was saying. I needed to hear her, even though our exchange was painfully one-sided. All I could hear were her baby grunts and breathing. Rosario said she did fine in our absence. She probably had no idea we were even gone. I, on the other hand, was heartbroken. As Julia got older, I realized that straining to hear her baby grunts wasn't nearly as painful as having to endure her cries:

"Don't go, Mommy...please don't go..."

But I had to go to work every day. And, more often than I would have liked, I had to travel out of the city to report stories. Sometimes, I didn't know how long I'd be gone.

When Gaby was born, Julia was just two and a half years old and not yet potty trained. Up to that point, if I had to travel on assignment, Eliott would take care of her and, on weekends, his two older daughters, Erica and Bianca. But two girls in diapers and two going through puberty was more than he could handle by himself, so we hired a weekend nanny— Ana Rosa to the rescue. She became my right-hand woman, helping me handle an increasing load of responsibilities. She didn't just help me with the children, she also helped me with my ailing mother, making sure she got her medication on time, that she ate well, that she got her afternoon stroll in the park.

My mother had a caregiver during the week, but on weekends I was in charge. Her condition deteriorated to the point where I had to ask my niece, Cici, to move to Miami from Los Angeles and help take care of her grandmother. We would alternate weekends, and eventually my oldest sister, Isabel, also moved to Miami and joined us in the rotation.

Before her stroke, I had always believed my mother would live well into her nineties. How could she not? In her late seventies she was vibrant and full of energy. She loved to travel and enjoyed her independence. But there she was, a stroke victim, frail, weak, and bedridden. I would call her every day and wait on the phone until, just like Julia had done as a baby, she would respond with a faint grunt or a heavy sigh. It was the circle of life, and my mother's was coming to its end.

One Sunday afternoon, as Eliott, my girls, my sisters, and I were lunching at the mall, I got a call from my mother's caretaker. She said my mom was wheezing and gasping. This had been happening for a while already, but I suspected this could be something else. To be safe, we decided to go and check on her. My sisters left first, and we told them we'd catch up with them at my mother's condo. But when I got there, I found my sisters in tears. Mami had stopped breathing.

Her doctors had warned us the end was near. And recently she had grown thin and frail, unable to eat, her body full of bedsores. I knew her quality of life had deteriorated to a point that no one should have to endure. My head told me I should be relieved that she was at last at rest. She was in a better place, no longer suffering. More important, as my Catholic upbringing taught me, she would be reunited with my father. But I was devastated. All those years of sharing a life, of watching over one another, nurturing an endless, unconditional love, had come to an end, and I wasn't there with her when she closed her eyes for the final time.

The hardest part of losing a loved one is learning how to live without them, how to get up each morning and not call them, how to come home after a grueling trip and not visit them. For as long as I could remember, my life had revolved around my mother. She had taken care of me until I started taking care of her. Now, more than ever, I felt I needed to follow her example.

But she had set a very high standard. My mother took her job as a seamstress as seriously as I take mine as a communicator. I'm fortunate to do what I do, but as a network anchor and correspondent, I don't own my own time. I am on call twenty-four hours a day, every day. At any moment my phone can ring with a news bulletin that could toss me into another time zone. At 2 A.M. one hot August morning in 1997 I got such a call from my news director at the time, Alina Falcón. Lady Diana had died after a terrible automobile crash in Paris. I had two hours to get on a flight to London to cover the reaction to her death. How long would I be there? Who knew? I thought at most I'd be gone three or four days, but the story kept me away for ten days. My baby Gaby was just three months old. Those ten days felt like ten weeks.

For the most part my daughters were oblivious to my traveling until they got old enough to realize that "goodbye" sometimes meant it would be days before they saw me again. And upon my return, I'd often have to go straight to work at the studio. After a while, instead of greeting me upon my return, my girls would hit me with a brutal question:

"Do you have work, Mom?"

It was so hard for them to understand why I had to leave all over again. But it was harder for me to leave them behind. As their pediatrician liked to reassure me, "The kids are going to be all right, but I don't know about the mom." I still get choked up when I have to say goodbye before going on assignment. I call them morning, noon, and night. I need to hear their voices and, more important, I don't want them to get used to my absence.

Like so many working moms, I often try to justify my dual role. My job is important because it allows me to forge a better future for my kids, provide them with the best education. Yes, we could do without the private school, the nanny, and the comfortable house. But do we want to? Do we need to?

I have tried to explain the importance of my job to my

daughters. I tell them I feel I can make a difference in people's lives, and they seem to understand. But there are the occasional not-so-indirect warnings that tell me to slow down. During one particularly hectic time when I was recording a regular radio commentary and writing a column for Univision's online portal in addition to my TV responsibilities, I'd come home, kiss my girls, and rush into my home office to write. I'd even work on weekends, racing to make my column's Monday deadlines. My oversized workload was not lost on Julia. One day in the car she told me about a strange dream she had the night before:

"There was this little piglet who was lost in the forest. She was really scared because she couldn't find her way home. She was crying really hard for her mommy. When she finally got home, her mom said, 'Okay, I'm glad you're home.' But the piglet was still really sad."

The end of the story caught me by surprise.

"Why was the piglet still sad?" I asked her.

"Because her mother wasn't excited to see her because she was too busy working on the Internet," she replied.

Talk about a dagger to the heart! Julia's message got through to me, loud and clear. I knew I had to make some changes in my work schedule. The first thing I did was change the due date on my column. I didn't want to spend any more weekends writing or doing the research. In fact, I try not to work at home at all anymore, unless the girls are at school or in bed.

I realized Julia's message was the same one my mother had taught me with her example. Children should not feel shortchanged by their parents' workload.

Gaby's jab was rather more unforgiving. Out of the blue one day, she declared:

"When I grow up I'm going to be a stay-at-home mom."

"Oh yeah?" I replied.

"Yeah, I'm going to show my kids I really love them. And I'm not going to have a nanny, either."

Her little eyes stared at me in a mean squint. She unnerved me to the point that I shot back a taunt of my own:

"Well, you won't be able to afford a nanny, because you won't have a job and, therefore, you won't have any money."

But she didn't stop.

"It's okay. My husband can work and support me," she came back.

I couldn't believe I was duking it out with a seven-year-old. I was sure she hadn't come to that theory on her own. She must have overheard some adult saying it at school. After all, most of her friends' moms are "stay-at-home" moms. I know those mothers work hard and end their days as exhausted as I do. I also know that, for all their little jabs—and the occasional body blow they land—my daughters are understanding and proud of me.

One time, Julia's first-grade class was asked to prepare messages for their mothers that described what made them special. The messages read as you would expect:

"My mother is special because she takes care of me when I'm sick."

"My mom is special because she tucks me into bed at night."

Julia's message read a little differently: "My mother is a respectable, responsible woman."

I might have preferred a more affectionate turn of phrase, but I found her message revealing. It told me she was proud of me, and that, perhaps, I was not only a mother, but a role model.

Nevertheless, many times I have struggled with the wisdom of my choices. True, most of my trips are not by choice, but there was one particular trip I did choose in the spring of 2001. It was during a momentous mayoral election in Los Angeles. Among the primary contenders was Antonio Villaraigosa, the former speaker of the State Assembly. This was one story I did not want to miss. Since the beginning of my career as a local reporter, I had been waiting for the day when the Latino community of Los Angeles would finally achieve the political representation they deserved. The empowerment of Latinos was an issue I had devoted a great deal of time and effort to as a journalist, and Villaraigosa had a good chance of becoming mayor. I wanted to be there to report on that historic milestone.

But I had a dilemma: go to L.A. and cover the election, or stay in Miami and attend Julia's kindergarten graduation.

I knew she wouldn't be alone. Her dad, sisters, aunts, grandparents, the whole family, would be there for her special day. Maybe she wouldn't feel my absence. I certainly hoped that was true when I flew to California to cover the story. And it was—she was just fine.

But to this day I have yet to recover from that decision. The guilt I feel is enormous. How could I go? Villaraigosa lost the election, but that wasn't the point. I had made the wrong choice. Villaraigosa could run again—and he did, winning the 2005 mayor's race in a landslide. In fact, he invited me to emcee his inauguration ceremony on July 1st of that year. But Julia graduated from kindergarten only once, and her mom was not there.

I try to mitigate the guilt by focusing on the quality of our time together, not on the days we spend apart. When I'm home, I make sure I'm the first person my daughters see when they open their eyes in the morning, and the last one they see when they close them at night.

On some nights, as I watch them drift to sleep, I feel as if I'm almost there, almost at that special notch of motherhood my mother showed me. I'm not sure I'll ever get there, but I feel fortunate to know it exists.

❧

TEN

Incense from a Distant Altar

My flight soared through a September morning en route to Mexico. It was a routine drill for me—pack up my laptop, my notes, my essentials, grab my passport, and go. I travel pretty well, and I actually look forward to the hours when I'm suspended in that capsule up in the sky, removed from daily distractions. I can think clearly up there, high above the clouds. But on this particular morning I was feeling a little nervous and out of sorts.

WHEN I am out reporting a story, I know what to do without hesitation: I research my subject, double-check my facts, strive for accuracy. I dig up details to enrich the story, and I always try to envision it from the perspective of my viewers. I want the story not only to inform them, but also to impact their lives. But this trip was a different kind of fact-finding mission. Whatever facts I uncovered in the next few days would not be for the benefit of my viewers. They were for me alone. My reporting would center not on any newsmaker but on my father. And when I closed my reporter's notebook—if I could close my notebook—there would be no script to write for the nightly news.

This was a self-imposed mission, one that I chose to carry out alone. My sisters hadn't shown much interest in exploring our father's past life. My oldest sister, Isabel, seemed intrigued when I shared whatever details I discovered; but she just listened and then went on with her life. She's a happy-go-lucky type of person who doesn't let herself get bogged down with needless preoccupations. My sister Tina, social butterfly that she is, met many of our relatives in Mexico years before I did. She accompanied my father in the last few years of his life to a couple of family reunions, where after decades he reconnected with sisters, nieces, and nephews. But she wasn't particularly close to him, nor had she expressed any interest in learning more about the decisions of his youth.

I felt alone up there. So many doubts and questions filled my head. What caused my father to pull away from his family? Had he felt ashamed in their eyes for leaving the priesthood? Had he been disowned for marrying a woman beneath his family's social class? Did they consider her unworthy of taking a name as distinguished as Cordero Salinas? But how could they not love my mother? She may have been poor, yes, but she was loving and generous. Maybe if they had taken the time to get to know her they would have accepted her into their aristocratic fold. Or perhaps the social divide never had anything to do with the family separation?

Whatever the reason for it, their estrangement colored my entire childhood, even if I didn't realize it until many years later. I grew up largely without aunts or cousins from my father's side, lacking that robust plural sense of Mexican families. Corderos. Salinases. Cousins. I had my sisters and maternal relatives with which to compare the quirks and traits of heredity. But did anyone else, I wondered, have my eyes, my hair, my temperament?

All those family truths awaited in the vast, arcane capital below. As the plane circled Mexico City in its final descent, I closed my eyes and sent up a prayer.

I GUESS I wasn't too hard to find. He knew my face from TV. The stout, business-suited man made a beeline for me in the hotel lobby. He walked with a sense of purpose, carrying a portfolio that bristled with papers and photographs. Meet Federico Jiménez Canet, my father's nephew.

Moments later we were sharing stories and photos at the hotel café with his wife, a friendly, chatty woman, and their two daughters. From his stack of documents Federico pulled out a large, impressive art book titled *El Pintor Juan Cordero*. It was filled with richly colored portraits and references to the great art collections of Mexico. Clearly, Juan Cordero was a revered master of his era. I knew that was my great-grandfather's name, but I didn't know that my great-grandfather was one of the most important Mexican muralists of the nineteenth century.

Among his most important works were portraits of the feared nineteenth-century Mexican ruler General Antonio López de Santa Anna and his wife, Mrs. Dolores Tosta de Santa Anna.

"One of his paintings is in the museum behind the Basilica of Our Lady of Guadalupe," Federico volunteered. "Another of his masterpieces, depicting Columbus before the Spanish monarchs, is on display at Bellas Artes, the most important performing-arts center in Mexico City," he said.

Through Federico I also found out that we have a second cousin, my father's niece, who sits on the Supreme Court of Mexico. Her name is Olga María Sánchez Cordero. Certainly the fact that we hailed from such illustrious stock was a source of considerable pride for Federico. Yet I detected something deeper than mere name dropping or empty boasting in his stories. He is a family historian of sorts, a man in love with the rich, compelling details of his ancestors' lives.

Between sips of coffee he shared stories of the family's struggle during the Mexican Revolution, how my father's father, José Antonio Cordero y Osio, a "rich junior" of the era, lost everything after the family's lands were taken away by the dictator du jour. He tried to survive as a photographer, utilizing a novel technique he had learned during his travels in Italy, but he was forced by circumstances to parcel out the children. Federico's mother, María de Los Angeles, was taken in by a convent as a cook. My father and his younger brother, José Antonio, were taken in by an older uncle, a priest named Rafael Salinas y Rivera, who at the time was in charge of the Basilica of Our Lady of Guadalupe. Father Salinas, beloved by the family, became my father's mentor.

"Why did my father leave the Church?" I asked, cutting to the chase.

Federico shook his head. "I don't know. When I was a boy, he'd come to my house. I always saw him as a very patient man, a man of character, a very cultured man," he said.

We lingered over a stately portrait taken in 1931 of the extended Cordero/Salinas family. In the faces on the vintage photograph, I traced familiar profiles, eyes, noses, hair. So many relatives. I recognized the family likeness in the faces of Federico's daughters. My cousins' smiles grew bigger when I showed them pictures of my Julia and Gaby.

At the end of our visit, I didn't seem any closer to finding out the reasons why my father had left the Church, but Federico's stories introduced me to a whole new Mexico of Corderos and Salinases. Cousins. And nieces and nephews.

LUCY ROMERO was one of those cousins who just surfaced in my life entirely out of context. An esteemed psychotherapist from Mexico City, she had a recurring guest spot on *Sábado Gigante,* Univision's popular Saturday-night variety show. While at our Miami studio one day, she ventured into the newsroom and asked to see me. She told me we were related. Her grandmother was my father's sister. It was Lucy who would put me in touch with the rest of the family, in particular with my father's nieces. She seemed well-informed and eager to guide me in my historical search. She had warm eyes, a firm handshake, and a serene, trustworthy sense about her. I felt as if I could talk with her for hours.

She was second on my list of appointments during my fact-finding visit to Mexico City. When I arrived at her counseling office, a cozy space inside an "interreligious center," she was busy with patients. While I waited, I studied the photos and bulletin-board items on the wall, a patchwork of images connected by a spiritual theme. Amid the photos, clippings, and announcements of the Dalai Lama's impending visit, one face stood out from the rest. Although she wore the tunic and headdress of a Sufi priestess, I recognized her childlike grin. It was Edlín Ortíz, a former Univision correspondent. She had transformed into Amina Al Jerrahi, a local leader of this mystical Muslim sect. And my cousin Lucy, raised by a fervently Catholic mother who in one memorable instance joined a Franciscan order as a laywoman, had embraced the Sufi philosophy.

"This is going to be interesting," I told myself as I stepped into her shabby-chic therapy room.

Indeed, it was. A handsome woman with blondish curls, Lucy recounted her family memories with the precision of an insider and the dispassion of a social analyst. Our family history, she said, was deeply interwoven with that of Mexico itself. She spoke of our ancestors in a larger context, noting the quirks of the upper-crust families of yesteryear.

"In every good society family, it was de rigueur to have a priest or a nun," she said, attempting to explain why both my father and his brother became priests. To hear her speak, it seemed as if my father's vocation was simply a consequence of his upbringing and greatly influenced by his uncle, Father Salinas.

"Cordero was a weighty last name, but he took the name Salinas in honor of his mentor," she concluded, adding that they had come together in the crucial years of the Mexican Revolution. My father's ministry, therefore, was forged in lean, chaotic times.

The priest's house and its many rooms surrounding a central courtyard was a shelter for the lost souls of the revolution, especially for those, like my father and uncle, whose parents were left penniless by rebel activity.

"Your father was a mystery," she went on. "He was one of the first persons I know of to hang up his priestly robes. He was an avant-garde man from a maverick generation, a man who dared to do what no one had the guts to do at that time. Of course, the price he had to pay was virtual expatriation and anonymity. The monolith of the Catholic Church has its own ways of erasing its errors. Why did your father expatriate? To erase his story and write a new one in the land of promise. And for the rest of his life he would carry that burden, that separation from his loved ones."

In one priceless nugget, there was the story of my father's journey through the priesthood. But it didn't answer my fundamental question: Why did he leave it?

I hugged Lucy goodbye. There we were, the daughter of a former priest and the daughter of a virtual nun, contemplating matters of family dogma in a kind of spiritual speakeasy. I left her in the prayer quarters, in the company of Edlín/Amina and various other believers in loose cotton garments. Together they hummed a prayer for me, wishing me success in my mission. Their hypnotic meditations mingled with the aromas of vegetarian cooking and drifted into the darkened street.

IF LUCY seemed well informed, then her mother, Lucila, was Information Central. We met for lunch the following day at the home of our cousin, Martita Palafox, one of the relatives I had met during my trip to interview Zedillo a few years earlier. Lucila took the bus from her home in the nearby town of Cuernavaca to bring me family stories, photographs, and relics. At first sight it was clear where Lucy got her regal looks, for Lucila carried herself with the grace of a vintage movie star.

Within the portrait-lined walls of Martita's house it was easy to feel at home. There were smiling faces of children at family gatherings, couples at weddings, cousins and second cousins, captured in all-too-familiar rites of passage. And there were papal proclamations, mementos from the Vatican, all interspersed with Mexican touches. But it took me a while to digest the fact that these women, twenty-plus years my senior, were my cousins, not my aunts. I tried to keep the branches of the family tree straight in my head:

Lucila is my father's niece. Daughter of María de Los Angeles. Federico's sister.

Martita is also my father's niece. Daughter of María Elena, "La Bebita," the aunt I knew best throughout my childhood, and La Muñeca's sister.

A gracious hostess, Martita lavished us with hors d'oeuvres and aged tequila, and later with a lunch of chipotle-scented vegetable soup and fish. We spent hours poring over photo albums as the cousins took turns offering snippets of family trivia. Of course, the art of Juan Cordero was well represented in the bottomless stack of documents. There was a slim Juan Cordero catalogue in which my father's brother, José Antonio, the more sociable of the two priests, had scribbled a dedication to the Mexican movie star María Félix.

Then Lucila handed me a small, velvet antique pouch lined with satin and embroidered with threads of gold, a relic so detailed and lovely that it looked like it had belonged to royalty.

"Your father used this to carry the Holy Communion Host," she explained.

The handmade pouch felt weightless in my hands. It transported me to another time and place, to darkened altars shrouded in veils of incense.

The cousins told me my father's last parish was San Juan de las Huertas, a small town on the outskirts of the capital. That must have been the small-town church the Opus Dei priest had refused to name.

My father had joined the "Felipenses," the San Felipe Neri

order of priests. Lucila described him vividly as a young Felipense priest serving at La Profesa, a venerable parish in downtown Mexico City. To hear her speak, it is clear her dear "Tío
José Luis" influenced her greatly in matters of the spirit.

"My uncle José Luis prepared me for my First Holy Communion at La Profesa," she recalled. She was very close to him,
she said, but he disappeared from her life when she was a
teenager. Lucila said the last time the family saw my father
was at her fifteenth birthday party. That was May 1, 1943. He
was in a very good mood, passing around refreshments and
chatting with all the guests, she recalled. The following day, he
vanished from their lives. May 1, 1943. I took note of the date.
In my search for the truth, it would be an important clue to
solving the mystery that was my father's life.

Lucila told me her mother, María de Los Angeles, took
his disappearance especially hard. She was convinced her
brother had slipped away to some monastery where he was
living as a reclusive monk. What else could explain the pro-
found silence? She took it upon herself to look for him. On
her travels to Latin America, she would stop at remote
churches and monasteries, and ask the priests about him. She
searched in vain, until one day, when she wasn't even looking,
she bumped into him in the least-expected setting. It was at a
party at the Mexican Consulate in Los Angeles on a fifteenth
of September, the eve of Mexican Independence Day. The
consul general of Mexico in Los Angeles at the time intro-

duced her and her husband to another Mexican couple. Imagine her surprise when the diplomat said: "I want you to meet Mr. José Luis Salinas."

"My mother was stunned. She thought, 'My brother! How could it be?'" Lucila recalled.

My father, who was probably equally shocked to see his sister, introduced the woman standing next to him.

"This is my wife, Lucita."

Lucila said her mother was so affected by the encounter that, upon her return to Mexico, she told no one about it.

"She couldn't bear to," said Lucila, as she continued to leaf through her albums.

But eventually word got out. My uncle José Antonio, who remained a priest, traveled to Rome to seek a dispensation for my father for having left the priesthood. And, according to Lucila, he got it. She remembered seeing a letter once, something written in Latin, that contained those details. But where was that letter now? She could not remember. Maybe one of her daughters had stashed it away along with pictures, documents, and other family mementos, she said.

She wondered aloud if my father had changed his name from Cordero to Salinas to conceal his true identity. Perhaps he felt ashamed of having left the Church, she speculated. I found her assessment to be jarring and most likely off base. Ashamed? I don't believe my father was ashamed of his choices, particularly since they led him to become a loving family man. But hearing these cousins go on and on about religion and the Church and the holy sacraments, I understood why any ex-priest would stay away from such an orthodox bunch, especially in the 1940s.

During our chat, Lucila seemed to be searching for something in her albums. And suddenly she found it. She pulled out a small photograph and handed it to me. When I saw the black-and-white image, I was taken aback. It was my father as a young man, dressed as a priest. Standing next to him was his brother José Antonio and Lucila as a young girl. By then I had already heard about my father's priestly past. I had conjured up mental images of him walking around in his cassock. But here, for the first time, was an actual photograph of the Reverend José Luis Cordero Salinas. I held it in my hands and stared at it for several minutes. It was my Papi.

Thanks to an afternoon with the ladies, listening to their stories and family anecdotes, I was inching closer to my father's truth. But I still had many questions. And the biggest one was: "Why?" Why did he leave the Church? I knew that if I wanted to find the answer, I'd have to peer into the mind of an ex-priest.

THE FORMER Father Alberto Athié took a sip of tequila and sank into his seat in the dim bar at the edge of a plaza in the artsy neighborhood of Coyoacan. Couples snuggled nearby as lounge music mingled with cigarette smoke. To our waitress, he was just another guy having a cocktail. But I knew his story. In fact, I reported his story during a one-hour special on the mounting accusations of sexual abuse by Catholic priests. Athié had once held highly respected positions within the Catholic Church in Mexico. He had been an international coordinator for the Vatican's charity, Caritas. He had also served as a leader in the Church's commission for peace and reconciliation in the insurgent Chiapas region. But an unfortunate incident left him to star in an insurgency of his own. We recounted the disturbing details that evening.

It happened in 1994, when he took the confession of a dying man. The patient, a former priest who had been rector of a prestigious university in Mexico, told him he had been abused sexually by his superior priest while in the seminary decades earlier. It was the first Athié had heard of the incident. Father Athié spoke to the man of the delicate balance between forgiveness and justice, and how one thing does not preclude the other.

"Forgiveness does not mean that we give up our search for justice," he recalled telling the man.

"Then I will forgive," said the former priest, "but I want justice to be done."

At the funeral mass, Athié, speaking in general terms, told the congregation of the former priest's wish. At the end of the service, several other former seminarians approached him with similar stories about the same accused priest. They wanted to go public with their allegations. Athié counseled them to seek justice from within the Church's own hierarchy. But eventually they told their story to a journalist contact at the *Hartford Courant*. And then all heaven broke loose.

In February 1997, the paper reported that nine former seminarians alleged they had been abused by the same superior priest when they were boys from the ages of ten to sixteen. The following year, a canon law case was formally filed against the priest. The Vatican's powerful Congregation for the Doctrine of the Faith deemed the claim credible enough to merit an investigation. But the case just sat there. No depositions were ever taken. It was a complex situation. The priest in question happened to be the Reverend Marcial Maciel, founder of the conservative Legionaries of Christ order in Rome, an influential priest and one of the late Pope John Paul II's closest loyalists.

Athié was told in no uncertain terms to back off. But his conscience wouldn't let him abandon the alleged victims or the deathbed promise he had made the former priest, and for his refusal to give up the case, Athié felt the wrath of his bishops. One by one, he was relieved of his duties. Unrelenting, he took the matter all the way to the Vatican. But, to his surprise, he once again ran into a brick wall. He was forced to leave Mexico City, and eventually, when the Church cut off all support, he left the priesthood altogether. He now works as a layman with peasants in the Mexican countryside on agrarian reform and other social issues.

In December 2004, nearly eight years after the charges were first reported, the Vatican reopened the investigation into Father Maciel's alleged crimes. But by then, it was too late for Alberto Athié, whose integrity had been his downfall within the Church.

Although I didn't know the circumstances under which my father had left the Church, I sensed there might be parallels between his case and that of Father Athié. I knew my father to be a man of conscience and conviction. He opposed war on moral grounds. In his Box of Secrets I had discovered letters documenting his refusal to enter the military service. He advocated equality and justice for the poor and disenfranchised. In fact, on nights when he'd listen to Cuba's Radio Progreso on his shortwave radio, he'd lament that Marxism was a squandered ideology. So perhaps he, like Father Athié, had found himself in a quandary of conscience. After all, my mother had told me he left the priesthood due to a great disappointment. Could it have been something as horrible as what drove Athié to hang up his priestly robes?

I could only wonder as we chatted and nibbled on chile-dusted peanuts in the bar that night. I had spoken to Athié about my father a couple of years earlier. So when I decided that it was time for me to really start digging, I asked him for help. I knew he would understand and not judge my intentions. He provided me with names of church officials who might open up the archives of the archdiocese. I would explore that route the following day.

At the end of our evening I realized Athié had no answers for me, in spite of his good intentions. But listening to his story all over again left me uneasy. If my father had experienced a similar disillusionment, it must have been so much worse. If the present-day Church proves claustrophobic for a priest as dedicated and seasoned as Father Athié, what was the Church of the 1940s like for my father, a relative novice?

I thanked my friend as he got up to leave. He seemed tired. It was the end of an exhausting day for him, but he still had to make another stop to check on a social work case. He was still very attached to his ministry, I could see. I watched him walk away and disappear into the plaza bustling with artists, students, and tarot readers. Another good soul miles away from home.

THE CHURCH of La Profesa stands in baroque splendor on the corner of Calle Isabel la Católica and Madero in the historic center of Mexico City. Outside its fortress-like walls, downtown traffic churns by and pedestrians shuffle past storefronts and signs boasting the latest hits, fashions, and flavors. Outside church walls, city life renews itself every day, every hour, every minute. Inside, time stands still.

I walked through the cool, ornate church one morning in hopes of finding someone who might remember my father. This was one of the churches where he served, for it is a principal temple of the San Felipe Neri order of priests. Built in 1740, La Profesa is so grand that it feels like many churches in one. Ten altars rise from its checkered-tile floors, hoisting Madonnas and saints and angels in spectacular motion. Constellations of tiny gold *relíquias* frame each Virgin, each one every bit as glorious as the one before.

At a side altar, I stopped to watch a young priest say mass. He couldn't have been more than thirty years old, if that. The sight of him made me feel strange, uneasy. I imagined my father decades earlier, offering mass as this young man now did. Suddenly, I didn't feel as if I was in a church but rather a movie set, a place removed from my own reality. I felt my heart race and my stomach stir. I couldn't help but see this young priest as my father, the sweet, loving father I lived with for thirty-one years. It was him, but in another life and another dimension. I couldn't shake the feeling even as I walked away from the mass and toward the church sacristy. That presence I felt, my father's presence, overwhelmed me. I tried not to cry but I couldn't hold back my tears. They swept over me as I passed all those glass-encased saints, seeking a sign of solidarity in their melancholic faces.

I made my way to the corridors tucked behind the main church and spotted a sign for the church museum. I followed the signs upstairs to the museum, hoping to find photos or documentation. But it was closed.

I found the small church office and asked the secretary if I could speak to Father Luis Martín Cano, the father superior of the local Felipenses. It was an urgent and personal matter, I told her. He was one of the contacts Alberto Athié had given me. She told me the pastor was not in but that we could contact him by phone. I felt strange as I took the receiver to speak to him. There I was, telling yet another priest the story of my father. As it turned out, Father Cano recalled not one but two priests named Cordero. But he didn't remember much. If I wanted more information, I would have to speak to Father Luis Ávila Blancas, an elderly priest who had been at the church for many years. He was the historian of La Profesa. But as my luck would have it, Father Ávila Blancas was out sick, convalescing from recent surgery. He would not be available for several weeks. I was leaving Mexico in two days.

As I left La Profesa, I felt as if I was leaving behind a secret, something sacred that belonged to me. I picked up a pamphlet of the church's museum. It described each room and contained a list of portraits. Among them was an "A. Cordero." Could it be my uncle? But why wasn't my father's name listed? Could it be he did not stay long enough? Or had they erased him from church history?

What had happened in this church? The great disappointment my mother spoke of—could it have happened here? I was so close to the truth, but I couldn't see it. I would have to wait for the answers.

AT THE suggestion of Father Cano, I made my way to the cathedral nearby. I crossed El Zócalo, the Constitutional Square, which joins the cathedral and the National Palace. The plaza was still festooned in garlands the color of the Mexican flag, in honor of Independence Day, and it brought to mind the image of my father's sister bumping into him at that consulate party.

This is my wife ...

How surreal for all of them. My father had gone out of his way to protect his private life, and here I was marching toward the church archives, ready to dig up his past. But I felt it was my right to do so. He was my father. That gave me the moral authority to ask, and the right to an answer.

As it turned out, the archdioceses archives were not at the cathedral. The attendant there sent me across town to the chancery, where I waited to see the priest in charge of the records.

Father Antonio Venegas was friendly and helpful. He allowed me to go through a thick book that documented the names of priests who had been ordained in Mexico City and the dates of their ordinations. It would help me a great deal to know when my father was ordained a priest. I tried to figure out the years based on my father's age. I was told you have to be at least twenty-four years old to become a priest. If my father was born in 1909, then his ordination would likely have been sometime around 1933 or 1934.

My father's name wasn't in the book. I still had no idea when he entered the priesthood, let alone when and why he left it. One of the many possible theories I considered to explain the absence of his name had to do with the government's persecution of the clergy. During the so-called Guerra de Los Cristeros of the 1920s, my father would have been in his teens, too young to be a priest; but the persecution resumed in the mid-1930s. Churches were closed, bishops were exiled, and priests went into hiding.

The chancery had very little information. Father Venegas told me that although my father had been a priest, it was most likely that his records remained with his order, not with the archdiocese. The priest must have noted my disappointment because he pulled up a chair and sat with me for a bit. My best bet, he counseled, was to write a letter to Father Ávila Blancas at La Profesa.

"Tell him you'd like to speak to him about a very personal matter and that you'd prefer to communicate in person," he said. Then he told me a story about a lay friend of his who was having marital problems. Although there were many factors contributing to his looming divorce, it was difficult to isolate the one, specific reason why he would choose to simply walk away.

"Those things are so difficult to understand," he said. "We can guess, but ultimately we may never know what goes on in another person's mind."

I appreciated his concern, but his words did not put my questions to rest. No, I couldn't possibly know what went on in my father's mind, but I could certainly find out what went on around him. And, more than ever, I was determined to do so.

ELEVEN

Undocumented Americans

> I, Luis Cordero Salinas, born in Mexico City on
> July 22nd, 1909, and a citizen of the Republic of
> Mexico ... hereby apply to Your Honor for reconsid-
> eration of reopening my case....
>
> *—undated letter to*
> *Attorney General Robert F. Kennedy*

My father may have left the priesthood, but it seems he never left behind his oath of poverty. It endured all throughout my childhood and into the rest of his years. But he was a wealthy man in other ways, in matters of the heart and the mind. He was an intellectual. He held several degrees. He could debate the fine points of philosophy, law, real estate, and baseball—in six languages. He found music in numbers, poetry in scriptures, humor in governmental policy. In my eyes, he was brilliant and complex.

In others' eyes, he was something else: a one-dimensional statistic. The worst of the bigots might have had two easy words to describe him: "illegal alien." I hate that phrase. Not only is it deeply offensive, it dismisses everything that matters about an individual and reduces him to a shred of paper. To the xenophobes, an immigrant is no different from a line of cocaine, a hot-wired vehicle, or a stolen handgun. He's illegal. End of story.

I grew up largely unaware of this level of prejudice. It wasn't until I began covering politics, City Hall, and the grass-roots efforts of the Hispanic community that I began to sense it. That's when my eyes opened. I realized the importance of my job, how I needed to be not only a journalist but a true communicator. I saw people in a community who were lost, who had no idea what their rights were. Many of them didn't speak English and were made to feel like second-class citizens, even if they were born here.

When I began to cover immigration issues, I realized just how deeply the intolerance toward this population runs. And when I realized that, it stung like nothing else. The epithets hurled by anti-immigrant types served no purpose other than to drive a wedge between "Them" and "Us." But what happens when that wedge divides a family? If I'm an American by birth, and my father is an undocumented Mexican immigrant, then what are we? Us or Them? It was clear these immigrant-bashers weren't in command of all the facts. These people are simply undocumented Americans.

When I started writing an English-language syndicated column for King Features some years ago, I explored issues of immigration policy from time to time. To my surprise, each mention of undocumented immigrants triggered an avalanche of reader mail, most of it hateful. After I wrote about the millions of undocumented immigrants who were being denied drivers' licenses in some states because they do not have Social Security numbers, I received enough mail to fill a California Department of Motor Vehicles office.

"Undocumented immigrants are crooks!" wrote one reader. "They are in this country illegally. They broke the law. Every one of them should be rounded up and deported. The undocumented immigrants you refer to don't need to drive. They need to be transported back to where they belong—the other side of the U.S.-Mexican border."

In a follow-up column, I noted my problem with this rationale: while it's true that some 8.7 million undocumented immigrants in the United States are Mexican, what would all the other "illegals"—the 1 million Europeans, the 1 million Asians, the 150,000 Canadians, the 624,000 South Americans, the 243,000 Africans, and the 16,000 Australians—do in Mexico?

Another reader insisted that Mexicans are inferior beings who come from a "lesser culture" and therefore should not be allowed to drive here. A lesser culture? If that's the case, then why is Mexico's music, art, and food enjoyed all over the world, most notably in the United States, where salsa outsells ketchup? The fact that I had to defend Mexican culture at all irked me. I mean, the column was about immigrants and driving, not the murals of Diego Rivera or the Aztecan origins of chocolate. But I quickly learned that the topic of immigration brings up all kinds of unfortunate tangents. I don't know where all that hatred comes from.

"Go back to your country!"

When people say things like that, I just have to wonder if I missed something on the news wires. Did California secede from the rest of the United States while I was out to lunch? Time and again I've had to explain that I'm a U.S. citizen by birth, in hopes they'd stop the *"go-back-to-Mexico-and-help-your-people-so-they-don't-steal-from-us"* nonsense. But it never seems to make a difference.

There had been a glimmer of hope once upon a time. In fact, as the governments of George W. Bush and Mexico's Vicente Fox warmed to one another in 2001, seeking solutions for chronic border concerns, there looked like there might be a logical, humane policy on the horizon. I interviewed President Bush at the White House on September 4, 2001, during the days when he and Fox were meeting to work out a migration accord. Bush seemed upbeat about the prospects for increased cooperation between his government and that of, as he put it, *mi amigo Presidente Vicente Fox.*

"We are working through the idea of helping people earn the ability to becoming legal citizens," Bush told me. "There are some interesting ideas of how people can take money from the U.S. that they would put in a retirement account and invest it in Mexico. That is an interesting concept that we are exploring, but we are in agreement that we've got to do a better job." I pressed him on the issue of legalization for undocumented workers, asking him several versions of the same question. But he would only say this: "It's part of a very complex issue. President Fox understood right off the bat that there would not be any blanket amnesty. That would not pass in Congress, nor do I support that. But what I do support is a process to enable workers to perhaps earn credit to come closer to a legalization process. That is all part of the migration issue that we are going to discuss, but I have to work with Congress, which has a big say in what happens."

Just days later, the border topic was no longer a matter of high priority. The meetings, the good intentions, the fragile accords all seemed to go up in smoke—literally. The terrorist attacks of September 11, 2001, saw to that. And even as the dust settled somewhat, nothing was the same. The deep-rooted hatred and mistrust of foreigners that has long plagued the American fringes resurfaced. Vigilantes took to the

U.S.–Mexico border, harassing undocumented immigrants and those of suspiciously Latino looks. The so-called Minutemen took it upon themselves to hunt down "illegals." For all their patriotic pretenses, they simply seemed to embody the worst of the paranoid xenophobes.

I wonder what my father would have thought if he had lived to see these Minutemen in action. Would he have been out there in the Arizona desert, raising placards in protest? I'm certain of one thing: he would have been as troubled by the treatment of his fellow immigrants as I am. It's truly disappointing to know that this is the way too many Americans view immigrants—as "illegals" to be hunted down. And it's even more disappointing when the rhetoric comes from an elected leader, especially one whose constituency includes a large number of immigrants. When former California governor Pete Wilson took up the immigrant-bashing cause, he thrust California into the headlines for all the wrong reasons. My native state is a proud trendsetter, but the kind of trend Governor Wilson started is not the kind that makes us all proud. With his Proposition 187, a sweeping measure that broadly denied basic services to undocumented immigrants, he sparked a national movement to put "illegals" in their place. The fury reached all the way to Congress, which in 1996 passed a series of reforms reducing federally funded services to *legal* immigrants.

When Wilson sought reelection in 1994, he went to town on the immigration theme. I remember one billboard in San Diego depicting immigrants sneaking across the border. The gist of its message: *They're invading us and we've got to take back our state!*

It was as if the immigrants were enemy invaders, alien creatures in a science-fiction movie swooping in from another planet to destroy our cities and our way of life. So outrageous was Prop 187 that, although it had passed overwhelmingly at the polls at the height of California's anti-immigrant fervor, it later was found to be unconstitutional in federal court.

I've had the opportunity to interview Pete Wilson twice in my career for Univision's network newscast. The first interview came at a particularly hostile time for immigrants in the United States. This interview, which we recorded in Los Angeles on May 14, 1994, started off on a contentious note. Earlier that day, we had taken a street poll of California Latinos. The question was: "If you had Pete Wilson in front of you, what would you ask him?"

On the street, they used to call him *"Señor Pik Wilson."* And they had a barrage of questions for him, mainly about immigration and social services. Needless to say, the governor didn't have many fans among the Latino population. When I began my interview, I described the mood on the street. As a transcript of our exchange shows, Governor Wilson didn't take too kindly to Latino opinions of his policies.

"There's a perception out there," I began the interview, "that you are a racist because of certain comments—"

"What comments?" Wilson interjected.

"Comments such as, 'Undocumented immigrants have the country in a state of siege, they're eroding our quality of life,'" I told him. "Many people consider your propositions to be inhumane and unconstitutional."

The governor was clearly miffed.

"Well, I completely disagree," he came back. "And I challenge anyone to present statistics to the contrary. When I say that two-thirds of the babies born in country hospitals of Los Angeles are born to parents who entered this country illegally, I'm stating a fact. That 40 percent of the births paid for by the state of California involve parents who arrived here illegally— that's a fact. That 14 percent of our prison population is made up of undocumented aliens who have committed felonies. They are costing the taxpayers of California 400 million dollars a year."

But I insisted that his language contained an unnecessarily racist and anti-immigrant tone. "You've said that undocumented immigrants have thrown the nation into a state of siege. Is it fair to make that comparison?" I asked.

"Well, I don't remember using that terminology," he replied.

"We have your remarks on tape, Governor," I told him.

"Frankly, it's not fair," he rebutted. "We're talking about illegal immigration. We are the most generous nation in history in terms of welcoming immigrants legally. All the other countries combined do not allow as much immigration as the United States does. But there is a massive amount of illegal immigration that probably exceeds the legal number. California gets half of the nation's immigration, so it does feel as if we're in a state of siege."

After trying to pin him down on statistics for a few rounds, I decided to toss him an easy question.

"Do you believe undocumented immigrants have contributed anything at all to the country?" I asked him.

He thought for a moment.

"I believe there are individuals who have contributed... This is a nation of immigrants, and we are proud of our heritage. That's not the point. I don't condemn them for trying to come here. I condemn the [Clinton] administration in Washington for making the problem worse."

At that point, we played a couple of questions from our street poll for the governor. He was clearly uncomfortable with the entire scenario. The final question hit a nerve.

QUESTION: "*Listen, Mr. Pik Wilson, I want to ask you if you would treat your immigrant grandparents the way you're treating the rest of us?*"

With that, Wilson got up to leave, cutting our interview short by a good ten minutes. He said he had a flight to catch, and he blurted his answer on his way out:

"Yes, I would. Because this is not racist."

THE SECOND time I interviewed Wilson was August 14, 1996, at the Republican convention in San Diego. I asked him again about his views of immigrants.

"There you go again," he said.

I took that to mean I had made an impression on him the first time around. As I read the transcripts now, more than a decade later, I realize that I'd hit the governor especially hard since I could not help but identify with the immigrant population he was targeting. Sure, he always attempted to make a distinction between "legal" and "illegal" immigrants, but I didn't buy it. Here was a guy who proposed that the U.S.-born children of undocumented immigrants be denied American citizenship. This is where he blurred his legal/illegal distinctions, where he really lost me. If it had been up to him, I wouldn't be a U.S. citizen. I would be one of Them because my father was one of Them. Whatever contributions I have made as a productive member of this society and as a dutiful taxpayer would have been lost.

And, like me, there are so many children of immigrants, documented or not, who were born here and who are generous contributors to our society. My mother had a green card, but it wouldn't have made a difference to Governor Wilson because she was married to an undocumented immigrant. Unlike the stereotypes the immigrant-bashers promote, my parents paid the hospital bill when I was born in Los Angeles.

This is an especially touchy subject for me. When I read some of the old letters from my father's Box of Secrets, I feel both frustrated and vindicated. Frustrated by the facts, vindicated by the eloquence with which he presents them. The letters document his long struggle to gain legal immigration status in this country. In a clear, firm hand, he told the story of how and why he crossed the border from Mexico. With meticulous detail, he described his trials and tribulations with what was then the Department of War.

My father's immigration problems began shortly after he arrived in the United States in the early 1940s. In a letter addressed to the Department of War one year later, he explained that he had come to this country to continue his "rigorous scientific investigation in the fields of sociology, philosophy, and history." He had landed a job with the National Schools, a chain of technical schools in Los Angeles, as a writer and editor for their publications on psychology. Before reporting to work, however, he was required to register with the army's local recruitment board.

In his letter, my father recounted that fifteen days after

having registered he was given a "1-A classification," as if he had specifically entered the country to join the U.S. Army. His efforts to appeal that classification failed and his request to be allowed to return to his country were denied. He believed he had been framed by the government.

In December of that year, he returned to Mexico without the required authorization. Ironically, while many undocumented Mexicans are accused of breaking the law when they cross the border into the United States, my father was considered a violator for crossing the border into Mexico, his own country.

"Considering that those who make up the Department of War are persons with sound criteria, as opposed to those in the Local Recruitment Board, whom, to my way of seeing things as a psychologist, are people with inadequate education, I would urge you to consider my case and absolve me of an offense I do not believe I have committed," he wrote. "I do this for my honor and because I am aware that agents of the Federal Bureau of Investigation are trying to locate me in Los Angeles."

He went on to list the reasons why he felt he was not obliged to serve in the U.S. Army:

1. I am a Mexican citizen by birth and by tradition, since all of my ancestors for the past three or four generations have been Mexican. As a consequence, because of my blood and my

culture, I am absolutely a NATIONALIST. My nationalism is mitigated by the conviction of promoting a sincere friendship with all the peoples of the world.

2. I am 34 years old and I have dedicated 28 of those years to my education. Jurisprudence, philosophy, history, sociology, art, religion, etc., have always been the object of my enthusiastic investigations. Therefore, I consider, without false modesty, that I have acquired an elevated concept of humanity, and, therefore, detest all acts of violence. I am determined to serve only the cause of civilization and culture within my own scientific field.

I found his third reason most striking:

3. Among many political ideologies, I believe I understand democracy and sympathize with everything that is acceptable in that system. However, placing it in a historical perspective, I have come to the realization that all political systems are subject to rectifications and, in some cases, contradictions. Therefore I refuse to sacrifice my life, which is the most esteemed human asset and protected by the law of nature, to defend a system that, in my judgment, lacks a seal of kindness.

My father was living in the Mexican border town of Tijuana at the time he wrote the letter. He was working as the manager of a radio station. He had no need to return to the United

States, he wrote, and offered to return his immigration permit to the local U.S. consulate if the Department of War would consider his requests:

1. *To kindly consider my case and absolve me of any wrongdoing, and to inform the Local Board of recruitment of your decision.*
2. *To extend a document that would allow me to save my honor and my personal security.*
3. *To know that you can count on me as a spontaneous and sincere collaborator in everything that relates to friendly relations between my country and our neighbor to the north, as well as any cultural or scientific activity pertaining to my expertise.*

Shortly thereafter, the Department of War responded, acknowledging receipt of his letter. "The matter has been referred to the Director of Selective Service in Washington, D.C.," it said.

He also received notice from the Coordinator of Inter-American Affairs: "Your case will receive every consideration by the authorities concerned."

Further complicating my father's case was the bizarre fact that upon leaving the United States in December 1943 he asked to be inducted into the Mexican Army. A desperate move, I imagine, especially from a man who abhorred violence. But judging from the content of subsequent letters, my father's struggle went on for several years, although the tone of his writing became significantly more conciliatory.

He returned to the United States, and apparently he was allowed to return with no hassle. But his immigration troubles were not over. An even bigger problem arose, this time threatening to separate the family. He was so desperate, he wrote yet another letter, this time to then–U.S. Attorney General Robert F. Kennedy. In the letter, he explained that a mix-up with an immigration agent at the border had resulted in another violation, which threatened his family's unity. He said he received a "border crossing card" from the agent based on an interview. But that card was considered to be fraudulent because, as my father wrote to Kennedy, it mistakenly described him as a legal resident. He was not. His residency had lapsed due to the length of his absence from the country.

"I went back to Mexico by voluntary deportation in 1952," he wrote. "If there is any provision in your legislation which provides means to avoid the deterioration of a family, protection for American-born children and forgiveness for the immigration laws offense that I committed, I beg your recommendation to get for the last time and forever the privilege of becoming a resident of the U.S.A. I am in the best of my strength and ability to work."

In another letter, he recounted how he was "excluded as inadmissible to the United States, under the provisions of

Section 212 (a) (19) of the Immigration and Nationality Act of 1952," and how he had since applied for permission to visit his wife.

> At the time that I made the application two of my daughters were born American citizens. Thereafter, another girl was born of my wife as an American citizen. They are registered at the Department of Health of Los Angeles. . . . The three girls were born at the Methodist Hospital, 2826 So. Hope St., L.A., Cal. The attendant was the same doctor, J. W. Whittaker.
>
> Admitting that I provided cause for my exclusion from the United States in accordance with the Immigration laws, I request now from your generosity and your sense of humanity clearly expressed in your Declaration of Independence and your Bill of Rights that you reconsider my case in order to give to me the special privilege to become again a legal resident of the U.S. as to provide for the moral education of my children, American citizens, and for their support.
>
> I am a Catholic. I have a complete university education and am healthy. Therefore, I am prepared enough and in the best of conditions to work and become a useful and active resident of the United States. . . .
>
> I declare that I am ready to do anything you request from me under the provisions of your laws in order to get the privilege of being admitted as a permanent resident of the U.S. and eliminate

the inconveniences of being separated from my family and to
properly attend to their complete education and their support.
     Waiting for your answer, I declare myself
     Sincerely Yours,

     L.C.S.

The status he was seeking did not come until decades later, just a couple of years before his death. Fortunately for his daughters, he didn't wait for it. He joined us in Los Angeles, where he worked, studied, paid taxes, and never spoke to us of his immigration problems. It wasn't until he was old and sick and yearning to see his family in Mexico that he asked us to help him become a legal resident. My handwritten notes dating back to this time, the early 1980s, attest to the application process. They, too, were in the Box of Secrets.

     "Left to avoid draft... Card picked up... Came back, 1962.
     Has been in U.S. since then."

Clearly he chose to live here without official status, but certainly not without documentation.

TWELVE

A Message from the Ruins

Of all the intangibles I inherited from my father, the one that has defined me most clearly as an American is his love of peace. When I recall his thoughtful phrases, and when I reread his 1944 letter to the U.S. Department of War, I realize that the apple didn't fall too far from the tree. I wholeheartedly share his conviction that national security depends not on the size of a nation's military budget but on its desire to seek nonviolent solutions to conflicts first and foremost. That doesn't mean I am incapable of balanced reporting in time of war. On the contrary, I believe true democracies are renewed each day by the freedoms they practice, and that includes a balanced press.

When I learned I would be sent to cover the war in Iraq in

the spring of 2003, I struggled with a mix of emotions. As a wife and mother of two young daughters, I knew my family would be worried sick about my safety. Sure, they were accustomed to my traveling on a moment's notice. But it's not quite the same to take off for the funeral of a beloved princess, a superpower summit, or an election in a Third World country as it is to go off to cover a war. It was a dangerous and unpredictable assignment.

As a journalist, I knew I had to go. This story topped the headlines I reported each night to my viewers, and would continue to do so for the foreseeable future. This was not the sort of story you cover by reading wire copy or watching daily Pentagon briefings. For this story, I needed to be on the ground.

Of course, getting on the ground, on Iraqi ground, was easier said than done, especially for journalists who were not embedded with U.S. troops. We had to rely on backdoor routes, hired translators, and private security guards, as well as on our own careful planning. But I don't want to get ahead of myself. Let me rewind a bit:

A few days before my trip, I came across the photograph of a young Iraqi boy in a news magazine. This boy had lost both of his arms and had been severely burned just days earlier when a U.S. rocket hit his home, killing sixteen members of his family, including his parents and his brother. Here was the poster child for collateral damage, certainly not the intended target of the war on terrorism, but a victim of it nevertheless. As I packed my bags for the flight to Kuwait, I couldn't shake the blank, painful stare of the little boy in the magazine.

A couple of days after I arrived in Kuwait City with cameramen Angel Matos and Herman Ulloa, and producer Margarita Rabin, I learned that the little boy whose dramatic story had made headlines around the world was still alive. He had been flown to Kuwait City, to a hospital that specialized in burn victims. His name was Ali Ismael Abbas.

We showed up at the hospital's burn unit, hoping to see the boy, and to interview him if possible. The chief surgeon who had been treating him told us the boy's condition was too delicate and that there had already been a media photo op the day before. The doctor told us Ali was responding well to treatment and that the skin grafts on his chest were adhering nicely. I told him how important Ali's story was to our viewers, many of whom had come from war-torn Latin American countries where children just like this boy suffered the consequences of war. We must have touched a nerve because he agreed to let us see Ali, but only for five minutes. We had to wear surgical masks and gowns to protect him from possible infection. We entered Ali's room to find an eleven-year-old boy transfixed before a tiny DVD player. What was he watching so intently? As I neared his bed, I realized this young victim of war was glued to a raging battle sequence between Japanese anime characters. Pokemon. One of his physicians, Dr. Sabreen Alzamel, agreed to translate for us during our brief visit.

"How is he feeling?" I asked the doctor. She turned to the boy and was clearly moved by his response.

"He wants his arms back," she said.

And he wanted his house and his father's car back, as well, Ali explained to the doctor. How else would he be able to work to support the five sisters who had miraculously survived the bombing? This was his message to the United States government.

Just a couple of weeks earlier, the world watched in astonishment as American troops toppled the statue of Saddam Hussein. It had become a central image of the war—the metal torso dragged through the streets by mobs of Iraqis celebrating the fall of the despised "Butcher of Baghdad." But here was a boy whose only crime was to be at the wrong place at the wrong time. He, too, was a symbol of this war, his torso maimed by the bombs intended to free the world of a ruthless dictator.

Ali spoke with passion and an intensity that transcended his years. His strength and resolve amazed me. I wondered what his village was like. Where were his sisters? How were they enduring the aftermath of so much loss? I was determined to get to Baghdad to find out.

Amid the bustle of doctors and journalists at the hospital, we overheard one man speaking to reporters in fluent, British-accented English. His name was Stewart Innes and he had come from Baghdad, where he worked as a translator for an Australian journalist who was covering the boy's story. He told us that the reaction in Australia to Ali's plight had been so overwhelming that one generous reader had offered to buy the boy a new house, or anything else he needed. What the boy needed was a better hospital, for at first he was being treated at a bare-bones medical facility in Baghdad's Saddr City slum. So Innes and the Australian crew convinced the U.S. military to transport the boy in a Black Hawk medivac chopper to one of the region's top burn centers in Kuwait.

Innes had come into our lives at just the right time. The Kuwaiti translator we had been working with could not travel into Iraq, so we asked Innes if he would accompany us. Luckily for us, he was willing and able to do it. He knew Baghdad very well, and, more important, he knew the way to the ruins of Ali's house.

MY CREW and I spent a full week in Kuwait, trying to convince U.S. Army officials to help us cross the border. We had thought we could hitch a ride aboard a U.S. military plane into Baghdad. It wasn't such a crazy thought—days earlier, U.S. military officials had extended the courtesy to CNN's

Christiane Amanpour and a few other high-profile journalists. But several days of lobbying the military brought us no closer to Baghdad. We even tried to convince them to let us tag along behind their tanks. That didn't work either. Instead, we found ourselves braving the mother of all sandstorms en route to Camp Victory, in Kuwait.

At the dusty camp, military officials had gathered a group of Hispanic soldiers for us to interview. These were young men, some no older than eighteen, from the Third Armored Cavalry Regiment out of Colorado Springs, Colorado, the oldest cavalry division in the United States. We met Sergeant Arturo Loredo, a proud young man whose Mexican father hoisted an American flag in the front lawn the day he became an army officer. And Alvaro Razo, a nineteen-year-old soldier from Michoacan, Mexico, by way of Fresno, California. He was living out his childhood dream. Not alone, mind you. He brought along his Virgin of Guadalupe prayer card and his rosary, which hung from his neck along with his dog tags. We also met Jesus Ortega of San Diego, California, and Israel Figueroa of Orlando, Florida, both twenty and both away from their families for the first time in their lives.

We set up our "videophone" and went live with their interviews on Univision's morning show, *Despierta América*. Through the magic of these satellite phones, we were also able to patch them through to their families in the United States.

As I watched them talk to their relatives, I saw not the leathery soldiers of a historic regiment but baby-faced boys, some younger than my twenty-one-year-old stepdaughter. They were apprehensive and homesick. We offered to take letters back home for them. Months later, we would hear from their grateful families.

WHILE OUR visit to Camp Victory proved to be worthwhile, it delayed our trip to Baghdad. I guess that was the idea, to distract us with "positive" stories and stall us, hoping we'd give up. I realized the U.S. military had no intention of helping us get to Iraq, either by ground or by air, after the chief communications officer spelled out his best-case scenario for us.

"You keep insisting that we fly you into Baghdad. So we'll fly you into Baghdad, but we won't let you out of the airport. You can get some shots from the airport and come back," he told me.

"No, we want to go to Baghdad," I insisted.

"Okay, then, we can drop you off and you can climb the fence and go into Baghdad. Then, when you get shot in the head, I don't have to be the one to call your husband and tell him," he retorted.

So much for our military escorts. Our "unilateral" passes

meant we were on our own. We also needed a border pass from the Kuwait government. But that one wasn't nearly as complicated to get. As we were filling out the paperwork, one of the Kuwaiti clerks caught a glimpse of the word "Univision" on our applications.

"Univision? Do you know Bert Delgado?" he asked.

Of course we knew Bert. He was a friend and a longtime employee of the network.

"His father was my professor!" he said. As luck would have it, this young Kuwaiti had not only studied in the States, but in our own backyard, at Florida International University. We had our passes in no time—thanks to Bert.

My three crew members and I joined a caravan of other "unilateral" journalists, translators, and a plainclothes British security guard on the twelve-hour journey through the desert into Baghdad. We took every precaution imaginable. We rented two four-wheel-drives, secured an extra set of tires, extra gasoline, the works.

After we finished our nightly news broadcast at 2:30 A.M. local time, we headed for the supermarket to stock up on supplies. We bought two generators, a tea kettle, water, Spam, tuna, crackers, cans of fruit, toilet paper, napkins, and tons of coffee. We packed up our bulletproof vests and gas masks, and at 4:30 A.M. we hit the road.

As dawn broke over the desert, I could see nothing but

sand for miles and miles. Along the isolated Iraqi road we spotted a few blown-up tanks, occasional rows of camels, and U.S. armored vehicles traveling in opposite directions. Nomads appeared from tiny specs of dust, their arms outstretched in a perplexing sign language. What did they want? Water? Food? Dollars? We had no idea. I gave one woman an apple and she just looked at it. She didn't speak. She just cried. What did she want? Who knows?

An hour outside Baghdad, the British security guard instructed us to put on our bulletproof vests. Soon enough, signs of a militarized town appeared. We made our way to the central zone blocked off by U.S. troops and found the Palestine Hotel, across the street from the plaza where the famed fallen statue of Saddam once stood. An informal headquarters for visiting correspondents, the hotel had been without electricity or running water for several days. Fortunately, a CNN producer, anticipating our arrival, had reserved two units for us in a small apartment building behind the hotel. So for our week in Baghdad, not only did we have ample room, but precious running water, cold but clean.

It was four weeks into the war. The city was virtually shut down. There were no traffic signals. No rules. Most government buildings had been blown up by U.S. "smart bombs," so smart that they had left the oil ministry untouched. On the streets, TV cameras followed looters as they dug and rifled through the luxurious remnants of Saddam's palaces. But we headed out to report on a different kind of excavation. Our translator was told a mass grave had been discovered. We followed his source's directions to the scene by way of a complicated route that required weaving through traffic jams and sweet-talking soldiers for access to certain roads. And as we finally made our way to the alleged grave site, we noticed a car was following us. We stopped to ask what they wanted. After a few tense moments, we asked them if they knew anything about mass graves in the area. They did. They pulled in front of us and we followed them to an abandoned area that had once served as a training center for the Iraqi secret police. Inside, as the road became more isolated and turned into a dirt path, we grew tense, wondering where they were taking us. But then we spotted a gathering of people standing before a long trench and mounds of dirt. This is where they were digging up bodies, hoping to find lost loved ones. As our camera rolled, they unearthed several cadavers.

Who had buried these bodies? Not even the people who were digging them up seemed to be sure. They could be victims of Saddam Hussein's brutality or Iraqi fighters buried by the Americans. Whatever the case, the gravediggers had taken over what seemed to be a training center for Hussein's feared Republican Guards, and left an eerie scene, littered with spent grenades and shells.

This early experience would set the tone of our coverage. We tried our best to stay off the beaten track. We ventured past the U.S. military lines into Saddr City, a no-man's-land formerly known as Saddam City, a slum inhabited by armed Iraqi toughs who peddled stolen weapons and munitions in their market. For years the Shiite majority had suffered the oppression of Saddam Hussein, who prohibited them from practicing their religion openly. Soon after the invasion, they took over the neighborhood, renaming it Saddr City, in honor of their spiritual master, the late Imam Mohammed Saddr, martyred by Hussein.

As we walked around, a crowd gathered and clamored in the background as I taped my stand-up. According to Stewart Innes, our translator, they were complaining about the lack of food, water, electricity, and safety. But again the gulf between us was enormous. As they clamored louder, Innes suggested I should return to the car. I started back, but the residents continued to crowd around me. I was pretty nervous, but I tried to be as friendly as I could be. I smiled and tried to show compassion. When I reached the car, I caught a glimpse of a nearby rooftop. There was a sharpshooter aiming the barrel of his gun in my direction. My heart stopped. Fearing the worst, I simply waved at him and smiled. To my surprise, he waved back.

What idea could these Iraqi residents have about who we

were and why we were there? I'm sure they were as frightened and confused as we were.

Back at our apartment at night, we wrote by candlelight and had to run a generator to power our editing equipment. My favorite accessory became a small, trusty flashlight that illuminated our path to our rooftop "set" overlooking the mosque across the street and then back to our rooms, up six flights of stairs.

At our apartment we met a group of Spanish physicians from Medics of the World who were working as volunteers at a nearby hospital. We followed them as they treated a ward full of Iraqis who had lost limbs in the violence. The doctors had brought medication and some medical equipment from Spain because the hospital was dangerously low on supplies. We interviewed mothers who told us how their children became the victims of errant explosives. There was a little boy who picked up a stray grenade that blew up in his hands. One little girl had lost her leg. Her sister had died. Each room brought more tragic stories and portraits of victims, all as heartbreaking as the story of Ali.

Midweek, with Innes's help, we set out to find young Ali's house. It was in a poor neighborhood on the outskirts of Baghdad. As we drove through the desolate area, I wondered how, in the era of "smart bombs," this explosion had been so devastating to these civilians? One local theory was that the intended target of the March 30, 2003, bombing was a suspected Iraqi military vehicle parked in an open field less than a mile away. But if that was the case, the bomb missed and instead destroyed five houses and several families.

On the crumbled block where Ali's family had lived, I found traces of his interrupted childhood—his Koran, his schoolbooks, and few other humble possessions. I met his sisters, ages six through nineteen. They were staying with an aunt until they could rebuild their home. They barely spoke. They were still numb from what had happened. Onlookers said they had been dug out from rubble after the attack. They knew their brother had survived, but they had not spoken to him since the bombing.

With our satellite phone, we were able to call the hospital in Kuwait City and put Ali in touch with his sisters. When the girls heard their brother's voice, they cried and cried.

"Ali! Ali!"

They had few words. But no amount of explosives had destroyed the bond between them.

I carried that image of them, the weeping sisters of Ali Ismael Abbas, with me as we crossed the desert once again toward Kuwait. They had said very little, but their message was deafening. Even in that gulf between languages and recognizable signs, a desperate plea rose from the ruins of one family's home.

We are not the enemy.

THIRTEEN

❧

Bless Me, Father . . .

I thought a great deal about my father during my trip to Iraq. All those images of war-battered civilians rattled me to the core. I wondered what he would have thought at the sight of a mother whose legs had been blown off, weeping in a Baghdad hospital, her children at her side.

I guess I know what he would have thought. He would have shared my feelings of indignation. He would have condemned the violence. Then he would have whispered a prayer.

I know what he would have thought because I carry his words in my soul. I heard him speak his mind. I have read his protest letters. I have taken his convictions and made them my own, embracing this sacred inheritance of passion. Other things, the innermost thoughts that led to his critical, life-altering decisions, have not been as easy to grasp. It has taken me years of questions, research, and interviews to come to some semblance of a conclusion.

I returned home from my fact-finding mission to Mexico City in September 2004 with more questions than answers. But as my list of questions grew, so did my determination to keep digging. I followed up my pending matters as quickly as I could.

I wrote a letter to Father Luis Ávila Blancas, the church historian who had been out sick when I visited La Profesa a few months earlier:

Dear Father Ávila Blancas:

I hope this letter finds you in good health. First, let me introduce myself. My name is María Elena Salinas. I live in Miami, in the North American state of Florida. I'm a journalist by profession, but the purpose of this letter is personal. It involves a delicate matter that I would have loved to discuss with you during my recent trip to Mexico. But unfortunately in those days you were indisposed.

I appeal to you as historian for the Felipense order, in hopes that you can help me clear up some questions and doubts I have about one of the priests who served in your order during the 1930s. The name of this priest is José Luis Cordero Salinas. The reason I'm interested in learning about his activities at your church is because he was my father. It was not until after he passed away that I learned by chance that he had been a priest. Several years have passed and my doubts persist, and I feel a great need to know more about his past.

I know it would be practically impossible to know the reasons that led him to change the course of his life, marry, and have three children, of which I am the youngest. But I feel a need to know more about the circumstances surrounding his time in the priesthood. For example, I'd like to know when he was ordained, why he chose the Felipenses, and what labor he carried out in the Church.

The only thing I know is that his brother, José Antonio Cordero Salinas, was also a priest. Father Ávila, I ask you from the bottom of my heart to share with me whatever details you might have, no matter how small or insignificant they may seem. For me it would be of great sentimental value. I love my father and keep his memory on a pedestal. But I cannot rest until I clear up these mysteries of his past, for they form part of my family roots.

I thank you beforehand for whatever attention you can give to this case. If necessary, I can return to Mexico to speak to you in person.

Sincerely,
María Elena Salinas

After sending my letter, I waited a couple of weeks before telephoning Father Ávila at the church. He sounded happy to hear from me. He had a sweet, grandfatherly voice and spoke in the amplified tones of someone who is a little hard of hearing.

"Yes, of course I remember Father Cordero," he said. "He prepared me for my first communion. I was just a boy, but I remember him distinctly."

I asked him if I could come to Mexico and meet with him. "I have so many things to ask you," I told him.

Of course he would meet with me, he said. But he warned that he didn't have too much information to share. He had lost track of my father shortly after those childhood catechism classes. That didn't matter to me; I knew in my heart that whatever he could add was that much more than what I already had in my notebook.

I called my cousin Lucila to give her the good news that Father Ávila had agreed to see me and that I would soon be back in Mexico. But she had even better news for me. She had found another two photographs of my father as a priest, and one of her daughters had found an old letter written in Latin. She believed it was from the Holy See. She said she would send them by messenger to my Mexico City hotel, because she was not able to travel from her home in Cuernavaca that week.

I packed up a few of the documents from the Box of Secrets and the photo Lucila had given me on my first trip—the one of my father in his cassock—and flew to Mexico City to meet Father Ávila.

Shortly after I got to my hotel (January 2005) I received the package from Lucila. The photographs she had spoken of arrived in mint condition. The dedications on the back were elegantly handwritten in a familiar hand, signed "Father José Luis Cordero y Salinas." The letter, a carbon copy, was also in very good condition, though on delicate, onion-skin paper. I didn't understand what it said, but it appeared to be some kind of formal or legal memorandum. I hoped Father Ávila had kept up with his Latin and that he might be able to translate it for me.

On the day of our meeting, I awoke extra early to give myself plenty of Zen time to review my thoughts. For the umpteenth time, I contemplated my reasons for wanting to know so much about my father's life. I could explain it away as a matter of identity and leave it at that. But was my father's business really my business? Did I have the right to pry into an area he had kept so off-base? I've talked about this with my sister Isabel. I've asked her why she doesn't seem interested in my parents' past. Her reply is short and sweet: she loved my parents tremendously and she is content with the love they gave her.

"If they wanted us to know, they would have told us," she says.

She is right, in a way. But although I share her gratitude for our parents' love, I don't share her complacency. I don't think my father's life was a stagnant parenthesis in time, but a running stream from which his loved ones and his ancestors can continue to draw knowledge and inspiration. What better tribute can I give the man I most loved in the world than to embrace his life, all of his life? To me, it's like saying, "*Gracias, Papi.* I know you loved me too much to burden me with your problems, but I can carry them now."

AT EIGHTY years of age, Father Ávila was just as I had imagined him, a frail but smiling priest who hobbled around with the help of a cane. He bundled himself in a sweater against the winter chill as he led me to his stately office inside La Profesa. The old, darkened church and its sweet fragrance of so many years of so many offerings gave me the feeling that I was entering a confessional of sorts. If that was the case, I wanted to be the confessor. I asked Father Ávila if I could record our conversation. He agreed.

We spent the next hour talking about the old days at La Profesa. His memory for detail was impressive, considering he was recounting things that had happened when he was ten years old.

"My parents used to send me to catechism lessons here at

La Profesa. So this is where I learned everything, from making the sign of the cross to reciting the Hail Mary. Everything related to Christianity..." he began. "I would come every Saturday. Catechism lessons were from four to five in the evening. There were two sisters, Guadalupe and Carmen Murguía, who would give us the lessons. And at the end, Father José Luis Cordero would come from the sacristy and give us the explanation of what we had learned....I remember him very well. He was a very distinguished individual, thin but with a very strong presence. I do remember that. And I also remember that one day after giving us the explanation of the catechism lesson—it was a Saturday—he told us, 'I bid you farewell. I am being sent to Rome and I don't know when I will return, so I have to say goodbye, but there will be another priest here available to give you the explanations.'"

His recollections were encouraging. I wanted to know everything he could possibly remember.

"What year was that?" I asked.

"It was 1936," he replied. "Then Father Cordero disappeared and I never saw him again."

I told him I heard my father had suffered a disappointment, that he had clashed with a resident priest during his time there, and that this priest would send him from one parish to another, never allowing him to settle in any one parish for too long.

"They said my father would build up a parish and then they would send him to another, and then another," I told Father Ávila.

"Yes, it's true," he said. "While your father was here, the father superior must have had some influence. He was very . . . how should I say . . . energetic."

"What was his name?" I asked.

"Father José González Rivera," he said

"And what did Father González do?"

"Well, he liked to bother people," he said, rather coyly.

"How did he like to bother people?" I insisted.

"He would treat them in a very despotic way, and it's very difficult to live with a person who treats you that way. Very difficult," Father Ávila lamented. "He was the father superior. But Father Cordero—I don't know what title he held. It must have been an important one because he put together a very rich collection of art that we have here in the church gallery. He was the first one who distributed the paintings, a collection that had been stored in the choir chambers, unappreciated. And Father Cordero began to distribute them and put them up, to appreciate them. And he also built some living quarters—just above where we are now—in which he lived."

My father had established a museum at the parish? Father Ávila must have noticed my face light up with pride.

"Indeed. He was the first. It is no longer the same because

I remodeled the area. I made it different, but I know for sure that he was the first one who organized this place and hung the paintings. It's too bad we don't have the pictures of that era. Another priest has them and they are filed away. During the short time your father was here, he undertook many projects. He built the gallery. He built the living quarters for the father superior and for himself and for one other priest who was here at the time. Before that there were no living quarters. There is a very small annex, and he was able to adapt the area so that the priests who served here could also live here at the church."

I reached for an envelope I had brought with some photographs of my father.

"I don't know if you remember him," I said as I handed the photos to the elderly priest. "That one is dated 1933.... And this one is from 1939. They were just given to me."

Father Ávila studied the black-and-white images with great curiosity. His eyes lit up at the sight of my father's face.

"Oh yes, here he is! This is exactly as I knew him. I knew him exactly like this, and it coincides with the date, 1933," he remarked. "In this other picture, from 1939, he was no longer part of this congregation. He was a secular priest, reporting to the bishop of Mexico."

I still had a million questions.

"Tell me a little bit about the Felipenses, because I would

like to understand why he chose this order. Why this one?" I asked.

"It is a very gentle community in its way of living, of interacting with others. It's joyful, simple. And one important element is that you can have your own belongings, which is not permitted with other orders in which you must take a vow of poverty. Chastity, yes, because we are priests. But we don't have to be poor, as long as we do not accumulate much wealth and obey our superior. Here we don't have such a drastic way of life," he replied.

"So among the different orders, this would be one of the least strict, would you say?"

"Yes, we live in a simple manner, that's all."

"What does it mean when a priest hangs up his priestly robes?" I asked.

"Exactly that," Father Ávila replied.

"But he would continue being a priest?" I wanted to know.

Father Ávila nodded decisively. "Oh, yes," he said.

His answer jarred me.

"So I lived with a priest all my life?" I asked in horror.

"Yes, of course," he replied.

Yet I wondered if my father had had any choice in the matter. What if he had simply come to the rectory one day in the 1940s and told his superiors that he wanted to leave the priesthood because he met a woman and wanted to marry her? What would his superiors have said? I asked Father Ávila.

"Well, you know, after the Second Vatican Council they allowed many cases, many cases.... Toward the end of Vatican II, many priests would leave to get married, and the Church married them," he replied.

I asked Father Ávila to take a look at two documents written in Latin, the one Lucila had just sent from Cuernavaca and a shorter one I'd found in the Box of Secrets. The latter, dated May 11, 1943, turned out to be a kind of permit. He read the documents silently.

"This is to allow him to say mass," he explained after reading one of the documents. "Yes, José Luis, your father, he could say mass in the United States, in Los Angeles and San Diego. So in that year, 1943, he was still functioning as a priest.... It's right here, look, this is a license, or 'permission'— that's what it's called. The bishop allows the priest to say mass."

The permit he was referring to was from the then-archbishop of Mexico, Luis María Martínez Rodríguez. Therefore, Father Ávila explained, my father still answered to the bishop in 1943.

"He must have gone to the chancellery and told them, 'I want to go to the United States. Give me permission to stay there and celebrate mass and take confession.' And they gave it to him," he explained.

The other document, the letter, was more complex. Father Ávila translated it into Spanish for me, but haltingly, with long pauses. He seemed to struggle not only with the language but also with the content. I couldn't imagine why. He had translated the first document effortlessly, yet when he picked up the onion-skin letter, he shifted gears.

"It's ... from the Holy See .... It says José Luis ... he left the ministry by his own will, and then had got married in a civil ceremony and has procreated three daughters," Father Ávila began. But then he seemed stuck on one passage. "No, that is strange ... it says he was already married in a civil ceremony."

"What else does it say?" I insisted.

"'He lasted ten continuous years, and because he suffered many annoyances, aggravations or disappointments that upset him, he decided on his own free will to leave the ministry,'" Father Ávila read, stopping to give me his interpretation. "Yes, he must have had a lot of problems with the father superior here. I did. He was a priest with very old traditions. And let me tell you something—we could not move one thing from its place. ... Can you imagine that? So when Father Cordero made all those changes—hanging the pictures and organizing the gallery—he must have gotten a lot of grief from this priest.

The father superior did not like anything to be moved. I put up with much of the same, and I suffered when I changed everything to the way it is now. But it was only for a short time, because I arrived here in 1960 and he died in 1965, and then I didn't have to put up with him anymore."

Father Ávila laughed at the thought, then returned to the letter: "It says right here that your father was annoyed and upset."

But then he came across a passage that seemed particularly important.

"Look, here they are granting him a special grace because he has recognized his faults," he said, reading the passage aloud: "'Taking into consideration his spiritual health and that of his family, a grace by order of the Sacred Apostolic Penitentiary, to be absolved of his wrongdoing . . .'"

Father Ávila put the letter down and delivered his best verdict.

"This means he died within the Church. He died having reconciled with the Church. It says here he is asking to be absolved and that he not be censored for marrying by civil ceremony and be allowed a legitimate union within the sacrament of holy matrimony. It says that before he died, he was allowed to marry by the Church. He reconciled with the Church—who knows, some priest may even have gone to confess him," he said.

But I wanted the sound bite, the bottom line.

"So they forgave him?" I asked.

"Yes, they forgave him," he replied.

Father Ávila noted my father had been advised not to return to Mexico, particularly to the places close to his faithful, to avoid causing a scandal.

"But he was forgiven, and allowed to receive other sacraments. Therefore he died within the Church," Father Ávila concluded. "This document is very important. Everything should be clear now. This should be a relief for you."

I followed Father Ávila to the small gallery my father had founded upstairs.

"It doesn't look anything like it did before," he warned me. He explained he had remodeled the area my father had once made into his sleeping quarters. I wanted to see it anyway, although I felt horrible that he would have to climb two flights of stairs in his condition. Father Ávila was gracious and helpful, and he knew how important it was for me to be there, to see it, to once again feel my father's presence.

"This is where your father slept," he said, showing me around the first gallery. "And in this area there was a kitchen. On this side, the dining room."

I had come so close to the gallery on my first visit to La Profesa. Back then, I sensed a connection to whatever was behind those locked doors, and I was right. I couldn't have imagined it at the time, but a part of my father's history rested in those very walls. And someone remembered.

* * *

WHEN I returned to Miami, I still couldn't shake the questions. Father Ávila had told me that my father's story had a happy resolution. But there was something about the way he read the letter, something about the way he looked at me during those long pauses that made me wonder what the ellipses had omitted. Was there something he was not telling me? Was he simply an old man whose Latin was rusty? Or was he a wise man of the cloth who simply recognized a daughter's yearning for closure and wished to grant her a measure of serenity?

If that latter scenario was the case, then I sincerely appreciated his empathy. But I wanted a second opinion on the letter. I contacted two Miami priests I knew and faxed them copies.

The first one, Father Alberto Cutié, had a strikingly different translation to offer. The letter, he said, was not from the Holy See to my father. It was from my father to the Holy See. The letter, he said, spelled out his request for a pardon.

The second priest, Monsignor Tomas Marin, concurred:

"He's asking the Holy See to eliminate his obligation to the priesthood. He mentions his daughters and his desire to live a Christian life and be united with his wife in a legitimate matrimony. 'For the health of my soul I ask that I be allowed to return to the sacraments.'"

Then my father had *not* died at peace within the Church? Here was the request, but where was the response? I called my cousin Lucila in Cuernavaca. She was the one who gave me that letter, so she likely knew more about its origins. I wanted to tell her what the Miami priests had concluded.

And it was she who would put my mind at rest once again.

"Yes, it is a request," she told me.

The letter, she said, had been written by my uncle, José Antonio. The address listed on the bottom," Avenida de la Union #351. Apartamento No. 3. MEXICO, 14, D, F," was where my uncle once lived.

She told me it was José Antonio who had traveled to Rome to personally seek the pardon for my father.

"And he returned with the pardon. I remember because he told me," she insisted. "'And everything is now fixed.'"

But this pardon, she added, came with four basic conditions for my father:

1. He could not say mass.
2. In case of war, he had to officiate as a priest.
3. He could not abandon his wife and children.
4. And he could not live on Mexican soil (to avoid the sin of scandal).

It was very important to the family that he be pardoned.

"We are very religious. We could not bear to have one of our own excommunicated from the Church," she said.

I trusted Lucila's memory. To reaffirm her story, she asked me if I had ever seen my father receive Holy Communion.

"Yes, I did," I replied.

"Well, he could not have done that if he had not been absolved," she said.

While that settled the questions I had about the Latin documents, it still didn't answer my other questions: When and why had my father left the priesthood? Dates, anecdotes, and photographs jumbled in my mind. Certain details stood out as I reviewed my relatives' stories:

My cousin Lucila had told me that the last time she saw my father was on May 1, 1943, the day of her fifteenth birthday party. He had officiated mass with his brother, José Antonio, then chatted with guests at the reception.

"He was in such a good mood. . . . Then we never heard from him again."

She couldn't have been confused about the date. It was her birthday. And, obviously, she knows what year she turned fifteen. But there was something about that year, 1943, that nagged me.

I remembered one of the many letters in the Box of Secrets. It was one of the first he had written to the Department of War, explaining why he refused to be drafted. What was it about Lucila's fifteenth birthday that reminded me of that letter?

I rifled through the tattered archive until I found the letter. I looked at the date: August 4, 1944. That wasn't it. Then I began to read:

> *José Luis Cordero Salinas*
> *P.O. Box 430*
> *Tijuana, B.C. Mexico*
> *August 4, 1944*
>
> U.S. WAR DEPARTMENT
> WASHINGTON, D.C.
>
> *My dear Sirs:*
>
> *I kindly submit the following case for your consideration.*
> *In May of 1943 I came into the United States through the border at Nogales, Arizona, as a legal resident of the United States.*
> *My intention for moving from my own country, Mexico, to the neighbor in the north, was with no other purpose than to continue the rigorous scientific investigations in the fields of Sociology, Philosophy, and History that have been my specialty since my youth. . . .*

May 1943? He crossed the border the same month he disappeared from Mexico City. But wait a minute—what about that Latin document authorizing him to say mass and hear confession in Los Angeles and San Diego? That form is also dated May 1943. So what was my father when he crossed the border at Nogales, a priest or a scholar?

And when did my mother come into the picture? I had spoken to so many people I thought could shed some light on what happened during those transitional years in my father's life. Most of the people who knew my parents well had already passed away. My father's brother and all his sisters had died, as had all but two of my mother's siblings. Her younger brother, Rodolfo, the same uncle who walked me down the aisle when I married Eliott in Puerto Vallarta, was unfamiliar with the intricacies of my mother's past. He was just a boy when she got married.

But there was one more person I needed to talk to. She might possess the missing pieces of this puzzle. I called her Nina, which is short for *madrina*, or godmother. I remember Esperanza Viades as elegant and soft-spoken. She was godmother to both Isabel and me. But in reality she was more than that. She was family. She was my mother's dear friend from her youth. I have countless pictures of her and her husband, smiling with my parents when they were young couples. My sisters and I grew up with her daughters, Sandra and

Espie—they were like cousins. Time and distance kept us apart physically, but the bond that was created through her friendship with my mother still endured.

So when I called my Nina after a couple of years of not speaking to her, she choked up at the sound of my voice.

"I think about you all the time," she cried. "I miss you and your mother so much."

She was in her mid-eighties and hard of hearing. We spoke for a little while about her life, her family, and her ailments. Then I asked her if she could remember when and where she met my parents as a couple. Were they already married? If not, did you go to their wedding? Where was it? Did they tell you my father had been a priest?

Nina wasn't sure of many details. She said she may have met them in Mazatlan, in the region of my mother's home town. She remembered they went to Mexico City to prepare their documents so they could travel to the United States. She remembered going to their wedding. It was a civil ceremony. But was it in Los Angeles? Mexico City? Mazatlán? She couldn't recall.

"Did you know my father was a priest?" I insisted.

"Yes," she said. "I even saw him say mass once with his brother. But then he left the priesthood and he worked as an ecclesiastical lawyer."

A lawyer. My mother had said he was a lawyer when she met him.

Nina's description of events was sketchy at best. But I was fortunate that she shared her memories with me that day, just months before she would pass away in November 2005. She didn't remember places or conversations. But there was one detail she didn't even have to think about. When I asked her what year she met my parents, she didn't miss a beat.

"It was 1943," she said.

I called Isabel to tell her I had spoken to our Nina. We chatted about the things I had heard from her and Lucila. Isabel was happy to hear about them. But as I spoke, I sensed she was drifting off in thought. I didn't want to bog her down with too many details—I know she's not so obsessed with these matters as I am. So I started to wind up the conversation. Then, suddenly, she interrupted.

"María Elena, you know I have Mami's wedding ring, right? And I think there's an inscription on it," she said.

Of course! My sister wore my mother's ring on a gold chain around her neck.

"What does it say?" I demanded.

I waited for her to read the inscription.

"It's a date," she said.

"What's the date? What does it say?"

Isabel uttered three numbers that nearly knocked me off my feet:

"Four. Five. Forty-three."

That, of course, meant May 4, 1943. The date inscribed

on the wedding ring was just three days after Lucila's fifteenth birthday party, just three days after my father disappeared from his former life. And now I knew why. He left to be with my mother. I had searched all over Mexico City looking for this clue. I had interrogated priests and cousins. And all the while the answer was there, dangling from Isabel's necklace. The *why*. I guess I will never know if his decision to leave the Church came before he met my mother or if her beauty and gentle soul captivated him so that he realized his vocation was that of a family man, not a priest. Maybe my father wasn't running away from anything—perhaps he was running *to* something. He hadn't left his vocation for a shameful reason. He left for love.

Now I know this in my heart. Yet, perhaps out of the very stubbornness I inherited from my father, I can't seem to close my notebook. Even when I traveled to Rome in April 2005 to cover the death of Pope John Paul II, I carried with me the questions that still haunted me about my father's life. This was a trip I had packed and unpacked for over and over again, for weeks, as the pope's health weakened. My fascination and reverence for this story was uncommon for a woman undergoing a crisis of faith, as I was. I felt estranged from the Church and its dogmatic confines. The "baby-killer" speech I had been subjected to during the 2004 presidential elections still haunted me, I confess. To me, it symbolized a certain arrogance and hypocrisy I found disturbing. In fact, the more I thought about it the more I realized that one of the reasons I had remained a Catholic was guilt—what would my father think? Was being Catholic an absolute condition for that "moral education" he so wanted for us? However, knowing he had left the priesthood—regardless of the reason—gave me an entirely different perspective. Gradually, I felt myself stepping out of the shadow of what I had perceived my father's expectations to have been, and I began to forge my own relationship with my faith.

I also felt guilt because I knew in my heart that I did not agree with the Vatican on some basic issues, such as divorce, family planning for the Third World, the role of women in the Church, and the lack of punitive action against pedophile priests. But there was something about Pope John Paul II that I found compelling. I felt a special connection to him. Like so many of the faithful I had met during all those papal visits I covered over the years, I was drawn by his charisma. Like many of those pilgrims, I was not as motivated by the Church as I was by this pope. There was something about him. He reminded me of my father.

The night the pope died, at 9:37 P.M., Vatican time, I was on the air, from Rome, anchoring Univision's special report from the Santa Monica Seminary. It was a privileged location for our marathon coverage. Just behind where I sat for our broadcasts was the window of John Paul II's apartment. Below us, in the streets leading to St. Peter's Square, a solemn hush descended upon the multitudes as news of his death spread. And when it came time to transfer his body to St. Peter's Basilica the following day, the throngs applauded its passage. Just as the pope had climbed into his pope-mobile so many times to come closer to his flock across the world, the world was now coming to him.

As I saw the video images of John Paul II's body moving through the streets, I was overwhelmed by emotion. Out of the camera's view, I felt tears stream down my face. I couldn't help but think of my father, also a man of God. I remembered the way my father used to gently touch our heads, as if in a blessing. I realized that I had always known, on some level, that his faith ran deeper than I could ever imagine.

I felt an urgency to seek out more answers, considering that I was in the city where my father studied, where perhaps he had been ordained, where his priestly order originated. I asked one of the first priests I interviewed where I could find the Church of San Felipe Neri.

"It's just down the street," he said.

That only added to my frustration, since I was tethered to the anchor chair. "Down the street" seemed a hundred miles away. A few days into my trip, as I headed for an interview, I chatted with Tomas Munns, a young, well-mannered college student we had hired locally to drive our van. We were his very first clients, as it turned out, and he was studying foreign relations. As we got to know each other, I mentioned that I wanted to learn more about the San Felipe Neri order, since my father had been a priest. Tomas's eyes lit up.

"I go to San Felipe Neri. My parents have been members of the church for thirty years," the blue-eyed young man told me, explaining that he was active in the youth group. He said there was a priest at the church who was over eighty years old. His father would be happy to take me to the church, he said.

I finally got a chance to visit the church on my last day in Rome. I met Tomas's father, who accompanied me. It was a glorious church steeped in history. Unfortunately, the priest we were looking for had the day off. Still, I felt as I had at La Profesa, in Mexico City—closer to Papi.

"What a small world it is, Tomas," I told our driver, thanking him for having invited me to his family's church.

"Nothing in life is a coincidence," he replied.

Returning to that church in Rome is still on my wish list in my active notebook, alongside all those details I'd like to investigate: Where was my father ordained as a priest? Did the Department of War ever exonerate him of any wrongdoing? Were my parents ever married in the Church?

There are so many other mysteries surrounding my father's life and his enigmatic personality. Maybe that Opus Dei priest, the one who had told me I should let my father take his secrets to his grave, realized the inevitable. Those secrets, and God knows how many others that I've yet to uncover, surely rest with my father in his grave.

When I was a little girl, I wanted to be just like my mother. Yet I grew up to be just like my father—rigid, obstinate, thirsty for knowledge, persistent. I know I'll always want to know more about him. Maybe that's because I'm a reporter. Then again, maybe it's because the more I find out about my father, the more I learn about myself.

EPILOGUE

My Dearest Papi,

I have so much to tell you and so many questions to ask. Let me start by saying that I miss you and my mother terribly. Now, more than ever, I realize how important the role of a parent is in the lives of their children. And even now as an adult, there are moments when I feel like a vulnerable and unprotected little girl. I wish you were here to give me guidance and support, to hug me and console me in the turbulent moments of my life. I wish you were here to share my joys and my triumphs.

You know, sometimes I think that during my childhood and adolescence I didn't appreciate you enough. I did not take full advantage of all the potential love you had to offer. I wasted the opportunity to tap into that universe of information that lived within you, your intellect, your vast knowledge of life, history, and culture.

But you must admit that you wasted many opportunities as well, for you could have been closer to my sisters and me. There appeared to be an invisible barrier that kept us at a distance, never allowing us to penetrate your soul. You could have taught us so much and encouraged us to make the most of our intellectual potential.

I now understand your attitude was probably due to your own rigid upbringing in a strictly conservative family. But those times are long gone. How I wish I would have had the chance to show you that women are capable of excelling and contributing to society in roles beyond that of homemaker, wife, and mother.

I know you were proud of my professional achievements. Your pride showed through. And it was important to me because I wanted nothing more than to please you and prove to you that I was doing something productive with my life. Back then I covered the city of Los Angeles as a local reporter. These days I travel the world, covering important news events and being a witness to history.

Oh, how I wish I could pick up the phone and talk to you about all those things you discussed only with men. It was obvious to me, Papi—discussions about war, politics, economics, legal ethics, and the latest medical advancements—that was men's stuff. Women needn't concern themselves with such topics. It would mean so much to me to be able to share my experiences with you these days, to have an open debate about politics, to analyze together the new world order.

I always saw you as a great man, but you were also an enigma. There was a veil of mystery that surrounded your inner being, your actions, your decisions, your physical and emotional distance. If I knew then what I know now, I would have fought hard to tear down that veil and get closer to you.

You cannot imagine what I felt when your friend brought me the now infamous "Box of Secrets" that you kept hidden from us. Once I opened it and explored its contents, I understood so much more about you. I always wondered why we didn't have as much contact with your family in Mexico as we did with my mother's. Frankly, I thought they disowned us because you'd married a poor girl from a small town. I grew up feeling rejected, even humiliated, by your illustrious family, whom I never even met.

The fact is, I never needed them. I have fond memories of my childhood. I loved you and my mother and always felt fortunate to have our small but close-knit family. We didn't have much, yet I never felt that something was missing in our lives. But now that I have had the opportunity to meet some members of your family in Mexico, I realize we could have enjoyed a fuller, more extensive family life, with more aunts, uncles, cousins, and shared life experiences.

Once I sat down to read your letters and analyze the documents you kept well hidden in your secret file, I was able to understand you better. I can only imagine how much you suffered, trying to guard your secrets. It must have been so difficult for you to make those decisions that first distanced you from your family in Mexico, and later on from your own wife and daughters. I admire your lifelong struggle to reunite with your family and to be accepted in the American society.

Learning about your secrets sparked in me a desire to know more. To find your family—our family. I wanted to investigate the events that led you to make such drastic changes in your life. I thought that if I dug deep enough into your past, I would better understand why your life—our lives—turned out the way they did. I wanted to know how the complexities of your life, your philosophy and the decisions you made so long ago, contributed to making me into the woman I am today.

I hold no grudge against you for keeping us so distant from the truth. I would like to think you did it to protect us. That you wanted to keep us safe and to make sure problems from your past did not confuse us or interfere with the Christian moral values you held so dearly and wanted so desperately to instill in us.

I know you always struggled to make ends meet financially and were never a success at business. You left behind no great inheritance for us. But what you did leave was an indelible mark on my innermost being. I appreciate the sacrifices you made for *my mother, my sisters, and for me. I appreciate the principles and values you ingrained in us. Your firmness and discipline, I now realize, put me on the path I walk today. You awoke in me a social conscience that is my moral compass in my life and my career. You showed me by example that what matters most in life is family, being a good person, respecting yourself and others. That is your legacy, and it is more valuable than any amount of money in the world.*

What I have found out about you is fascinating. It has helped me understand you and understand myself better. It has made me love you more and realize how important you were in my life. I will always follow your example and advice. I will always hold your name up high with pride and dignity. I will try to teach my daughters what I learned from you. With my head held high, I will proudly say I am my father's daughter. But, Papi, I want you to know, that I want my own life to be an open book.

I love you.

Your daughter,

Malenita

February 14, 2005

ACKNOWLEDGMENTS

One of the most wonderful things about human relations is that you can be touched in so many ways and learn so much from different people who come into your life. I know I have. If I were to name them all I would need another book. So, for now, I would like to acknowledge those who helped make this book a reality and those who have had influence on my personal and professional growth.

This book was in the works for many years. It took on different forms. I wrote and rewrote several proposals, but it wasn't until René Alegría sat down and read one of them that it began to take shape. I thank René for his interest and enthusiasm. Still, being a multimedia journalist, housewife, and mother of two daughters and two step-daughters didn't give me much free time for an added responsibility. If it had not been for my literary agent, Bill Adler, I might not have gone through with it. "It's no longer your choice," Bill said. "You have to tell your story." Thanks for the push, Bill, and for simplifying the process. And thanks to Glenn Mott of King Features for introducing me to Bill, for believing in me as a syndicated columnist, and for supporting me as an author.

For over two decades my coanchor, Jorge Ramos, and I have been witnesses to some of history's most critical moments. Together we have watched and reported on the incredible growth and influence of our Hispanic community. We've shared both tense and intense moments, politically charged and emotionally shattering. And through it all he has always been graceful, respectful, and serene. As an author, Jorge has led the way by example and constantly encouraged me to write my story. Thanks, Ramitos, for being a great partner, for answering my million questions, and for listening to my insecurities.

Few people cheered me on to write the book like Sylvia Rosabal-Ley, even before she was my boss. I thank her for believing my story was compelling enough to become a book. I consider myself fortunate to have had so many great news directors. Before Sylvia, it was Alina Falcón whose intelligence, kindness, and fairness are an inspiration. Guillermo Martínez enlightened me about the delicate balance in covering politically sensitive news events, and how a working relationship can turn into a friendship. From Pete Moraga, my first news director, I learned not only the basics of journalism but how to apply them with humanity and humility. Thanks to Frank Pirozzi for supporting my new projects with a smile, and to Ray Rodríguez for his leadership and graciousness.

Our news department has enjoyed success for many years not only because of the leadership of our management but because we are a great team. I feel privileged to have worked with the professionals behind *Noticiero Univision*, in the newsroom and on the field. I thank Marilyn Strauss, Patsy Loris, and Lourdes Torres for sharing tragic and historic moments with me, both personally and journalistically. Angel Matos has been more than the eye behind the lens in my travels around the world: he has been my friend and protector, and for that I thank him.

Thanks to my special friends Teresa, Lazz, Emma, Roy, Manny, and especially to my beloved Regina for not allowing time and distance to get in the way of our friendship.

I thank Liz Balmaseda for her incredible talent as a writer but also for her sensitivity and compassion; for treating my story as if it were her own and my innermost feelings with the utmost respect. Thanks to my newly found cousins and aunts in Mexico for enlightening me about our family history, to my Nina Espie for her kindness, and to my cousin Frida for always being there for me.

No professional success can compare to the importance of family. Yet it's ironic how sometimes it is they who suffer most because of our time-consuming professional endeavors. I thank my family for their patience, support, unconditional love, and for helping me put my life into perspective.

Thanks to my daughters, Julia and Gaby, for being my dream come true and putting up with my workload. To my step-daughters, Bianca and Erica, for showing me how love and respect can be earned. To my niece, Cici; my nephew, Charlie; and my sisters, Isabel and Tina, for keeping together our family ties. And to my mother, Lucy, and my father, José Luis, thank you for being such wonderful human beings and the source of the values that I have built my life upon.